SpringerBriefs in Microbiology

More information about this series at http://www.springer.com/series/8911

Monika Glinkowska · Lidia Boss
Grzegorz Wegrzyn

DNA Replication Control in Microbial Cell Factories

 Springer

Monika Glinkowska
Lidia Boss
Grzegorz Wegrzyn
Department of Molecular Biology
University of Gdańsk
Gdańsk
Poland

ISSN 2191-5385   ISSN 2191-5393   (electronic)
ISBN 978-3-319-10532-1   ISBN 978-3-319-10533-8   (eBook)
DOI 10.1007/978-3-319-10533-8

Library of Congress Control Number: 2014950672

Springer Cham Heidelberg New York Dordrecht London

Printed on acid-free paper

Springer is part of Springer Science+Business Media (www.springer.com)

Contents

Contents

DNA Replication Control in Microbial Cell Factories

Abstract The process of amplification of genetic material is crucial for production of progeny cells and organisms. Any obstacles in completion of faithful genome replication threaten cellular integrity. Therefore, this process is tightly regulated, mostly at the initiation phase, and—particularly in bacterial cells whose environment may undergo fast changes—adjusted with current growth conditions. Studies of recent years provided evidence that regulation of DNA replication in bacteria is more complex than previously anticipated. Multiple layers of control seem to ensure coordination of this process with the increase of cellular mass and the division cycle. Metabolic processes and membrane composition may serve as points where integration of genome replication with growth conditions occurs. It is also likely that coupling of DNA synthesis with cellular metabolism may involve interactions of replication proteins with other macromolecular complexes, responsible for various cellular processes. Thus, the exact set of factors participating in triggering the replication initiation may differ depending on growth conditions. Therefore, understanding the regulation of DNA duplication requires placing this process in the context of our current knowledge on bacterial metabolism, as well as cellular and chromosomal structure. Moreover, in both *Escherichia coli* and eukaryotic cells, replication initiator proteins were shown to play other roles in addition to driving the assembly of replication complexes, which constitutes another, yet not sufficiently understood, layer of coordinating DNA replication with the cell cycle. In this work, we describe the current knowledge of biochemical mechanisms regulating initiation of DNA replication in *E. coli*, which focuses on the control of activity of the DnaA protein, and we expand the view by providing examples of direct linkages between DNA replication and other cellular processes. In addition, we point out similarities of the mechanisms of regulation of DNA replication operating in prokaryotic and eukaryotic cells, and suggest implications for understanding more complex processes, like carcinogenesis.

1

M. Glinkowska et al., *DNA Replication Control in Microbial Cell Factories*,
SpringerBriefs in Microbiology, DOI 10.1007/978-3-319-10533-8_1

Introduction

Development of in vivo imaging techniques during recent years uncovered a new world of bacterial cell structure, revealing previously unanticipated level of spatiotemporal control of its components (Govindarajan et al. 2012; Vandeville et al. 2009). Several of these findings will be highlighted in the following paragraphs, giving an overview of the major breakthroughs and providing references to a number of excellent reviews comprehensively presenting achievements in particular fields.

Results of special importance for the progress in understanding subcellular organization of bacteria came from studies on divisome as well as chemotaxis and motility-governing structures. They provided evidence on the formation of specialized and highly localized protein complexes, whose components form a densely connected web of interactions (Rudner and Losick 2010). Moreover, many other proteins have been shown to form clusters performing particular cellular functions, assembled according to external and internal stimuli. They often take a form of helical filaments (Ingerson-Mahar and Gitai 2012; Celler et al. 2013) or localized groups within the cytoplasm and cellular membrane (Shapiro et al. 2009). Recent studies performed in *Caulobacter crescentus* demonstrated that around 10 % of all proteins display non-random intracellular distribution (Werner et al. 2009). Mechanisms responsible for specific protein anchoring and their dynamics within the cell have started to be uncovered, and they include membrane heterogenic composition (Fishov and Norris 2013; Mileykovskaya and Dowhan 2009) and curvature (Ramamurthi et al. 2009; Ramamurthi and Losick 2009), interactions with cytoskeleton elements (Heichlinger et al. 2011; Kawai et al. 2009; White et al. 2010), and chromosomal confinement (Kuhlman and Cox 2012; Saberi and Emberly 2013).

In concert with non-random intracellular allocation of proteins, distribution of mRNA was also found to be spatially coordinated (Campos and Jacobs-Wagner 2013). Although little is known about mechanisms ruling mRNA localization, results obtained from *C. crescentus* and *Escherichia coli* show that dispersion of mRNAs from the sites where they were transcribed is limited (Montero-Llopis et al. 2010), which suggests that chromosome organization itself may provide a framework for increased concentration of encoded proteins at specific subcellular localizations.

In line with the above findings, the view of bacterial chromosome structure has also undergone a major transformation. Once thought a passive store of genetic information and a random entanglement of DNA strands, it is now recognized as a highly organized entity and the influential role of its structure on the regulation of DNA transactions (as well as the reciprocal determination) has been underlined (Dorman 2013; Fogg et al. 2012; Muskhelishvili and Travers 2013; Wang et al. 2013). Last but not least—membrane composition turned out to be heterogeneous, while its curvature was implicated in defining the localization of protein complexes (Ramamurthi and Losick 2009; Fishov and Norris 2012). Thus, recent discoveries have altered the perception of bacterial cell and uncovered new levels of organization of cellular processes. In this light, it seems surprising that current understanding of the regulation of DNA replication in *E. coli*, the crucial biological process, relies mostly on the description of its biochemistry—the activity and sequence of actions of the proteins involved in its consecutive steps.

Existing models provide little explanation for the coordination of the DNA replication with cellular metabolism. They are based on the fluctuations of the amount of active form of the replication initiator—DnaA, available for binding with the *origin* of replication. However, the exact explanation how these fluctuations are correlated with growth conditions remains elusive. The prevailing view is that in *E. coli*, accumulation of the replication initiator is coupled to cellular mass, and thus, DNA replication starts at a specific mass-to-*origin* ratio (initiation mass) (Donachie 1968). However, whether this ratio is constant or changes with growth rate remains still a matter of debate (Wold et al. 1994). Moreover, support for the cell mass-dependent replication mechanisms outside *E. coli* is limited (Hill et al. 2012). In addition, mechanism coordinating DNA replication, cell division, and maintenance of cell size homeostasis remains unclear. Here, we would like to emphasize that understanding of these issues calls for placing regulation of DNA replication and cell cycle control in the context of emerging view of the organization of bacterial cell and metabolism. In particular, we would like to point out the existence of a large body of evidence suggesting direct coordination of DNA replication in *E. coli* and *Bacillus subtilis* with major cellular functions: central carbon metabolism (CCM), transcription, translation, and chromosome dynamics. We suggest that this coordination may be achieved by communication between protein complexes engaged in these processes and realized via direct protein–protein interactions and possibly—also by changes in the DNA structure. We provide also examples of similar mechanisms operating in eukaryotic cells.

Coordinating Duplication of Genetic Material with Cell Growth: Current Views and Old Questions

Natural habitats of *E. coli* do not offer stable life conditions. Therefore, bacterial cells have to quickly adjust their physiology to changing environment and availability of nutritional resources. During feast, the buildup of cellular mass is faster and cells are bigger, whereas at famine, the growth is slowed down and cells are reduced in size. Before division, genetic material has to be duplicated in an energy-consuming process of DNA replication, and the chromosomes need to be segregated before septation. Sufficient resources must be available for all these processes to be successfully completed. Therefore, mass accumulation, chromosomal DNA replication, segregation, and cell division must be coordinated to ensure that the chromosome is duplicated only once per cell cycle, and each daughter cell inherits its faithful copy while maintaining proper size. Consequently, in bacteria, DNA replication, as well as synthesis of other key macromolecules—RNA and proteins—is coupled to nutrient availability and growth rate (Dennis and Bremer 1974). How cell cycle parameters are adjusted to account for changes in growth rate is a fundamental and outstanding problem in bacterial physiology. Current understanding of these issues draws mainly on the studies of Cooper and Helmstetter (1968), done over 45 years ago.

To describe the results published by Cooper and Helmstetter (1968), it is necessary to present cell cycle events on a timescale, where it can be dissected into three successive periods: the period between the end of cell division and initiation of DNA replication (B period), the time required for the synthesis of chromosomal DNA (C period), and the interval between completion of replication and the end of cell division (D period). In their work, Cooper and Helmstetter showed that under constant temperature conditions, an increase in nutrient availability is accompanied by a decrease in mass doubling time, whereas C and D periods' durations remain constant for the generation times lower than 60 min (Cooper and Helmstetter 1968). To explain this, they proposed that the faster the bacterial cells grow, the more frequently the DNA replication is initiated, and consequently, the next round of DNA replication starts, while the existing one is still ongoing, which results in multifork replication. This conclusion allowed us to clarify how bacterial population can grow with a doubling time much shorter than the one required to synthesize a full copy of the chromosomal DNA (Cooper and Helmstetter 1968).

The above model provided foundations for our understanding of DNA replication control in *E. coli*; however, it presented the cell cycle as a process consisting of consecutive steps, whose timing is set by the initiation of DNA replication. It was proven, however, that bacteria growing with a doubling time over 70 min adjust replication rate to metabolic conditions by modulating duration of the C period (Michelsen et al. 2003). Thus, the Cooper-Helmstetter model neither accounts for metabolic control of the elongation of DNA replication, for which the evidence has accumulated (Janniere et al. 2007; Maciag et al. 2011), nor it explains how the cell determines when is it the proper time to commence the DNA duplication process. As an alternative, it was proposed that the cell cycle consists of separate processes of cell growth, chromosome replication, segregation, and cell division, which are independently regulated but interrelated by checkpoints ensuring the proper order of events (Boye and Nordstroem 2003; Wang and Levin 2009). We suggest that metabolic pathways and communication of the replication machinery with other macromolecular complexes carrying out basic cellular functions could constitute such checkpoints. In the following sections, we summarize current view on the details of the regulation of DNA replication in *E. coli* and point out some of the outstanding questions regarding these mechanisms.

It is widely accepted that the initiation of chromosomal DNA replication in *E. coli* is set by the availability of the DnaA protein in its active, ATP-associated form. Overexpression of DnaA results in too frequent replication initiations, whereas its depletion delays or blocks the start of the next replication round (Atlung et al. 1987; Bremer and Churchward 1985).

Structural and biochemical analyses indicated that DnaA consists of four functional domains (for a review, see Kaguni 2011; Katayama et al. 2010, Ozaki and Katayama 2009). Domain I (amino acids 1–90) plays a role in oligomerization of DnaA (Simmons et al. 2003; Weigel et al. 1999) and DnaB helicase loading (Marszałek and Kaguni 1994; Sutton et al. 1998; Seitz et al. 2000; Abe et al. 2007) and mediates its interaction with several regulatory proteins (Chodavarapu et al. 2008 a, b). Domain II sequence is not strongly conserved among DnaA orthologs (Messer 2002), and its length rather than a particular sequence is important for the protein function (Nozaki and Ogawa 2008); hence, this domain was proposed to act as a flexible linker between the DnaA core (domains III and IV) and the domain I. Nevertheless, a role of domain II in the stimulation of the prereplication complex formation by the DiaA protein was implicated (Ishida et al. 2004), whereas the removal of amino acids 96–120 resulted in the suppression of over-replication phenotype of *seqA* mutants, proving that this part of DnaA is not dispensable for the coordinated protein activity (Molt et al. 2009). Domain III (aa 130–347) contains motifs characteristic for the AAA+ superfamily of proteins (Duderstadt and Berger 2008) and takes part in ATP binding and hydrolysis (Nishida et al. 2002; Erzberger et al. 2002; Kawakami et al. 2006), DnaA oligomerization (Felczak and Kaguni 2004; Erzberger et al. 2006, Kawakami et al. 2005), and DnaB binding (Marszałek and Kaguni 1994; Seitz et al. 2000). Domain IV contains helix–turn–helix motif and is responsible for specific recognition of the DnaA box (Fujikawa et al. 2003; Erzberger et al. 2002; Blaesing et al. 2000; Roth and Messer 1995). A region at the boundary to domain III forms a hinge which allows for a rotational flexibility of domain IV (Erzberger et al. 2002). Analysis of the structure of domain III/IV from *Aquifex aeolicus* revealed that DnaA–ATP assembles into a compact, right-handed helical filament (Erzberger et al. 2006). DnaA associated with ADP lacks this oligomer-forming property (Erzberger et al. 2002), and thus, complex formation by DnaA is regulated via the nucleotide switch.

It has become increasingly evident that the information about assembly of the DnaA oligomer competent for the initiation of DNA replication is encoded in the structure of the replication *origin* (for a review, see Leonard and Grimwade 2010, 2011). Minimal 245-bp *oriC*, capable of autonomous replication in *E. coli* (Tabata et al. 1983), comprises of two functionally distinct parts—DnaA *a*ssembly *r*egion (DAR), where DnaA oligomer is formed, and *D*NA *u*nwinding *r*egion (DUE), where DNA strands become separated during the initiation to enable helicase loading. In early studies, comparative sequence analysis of this region in *E. coli* and a few other enterobacterial species allowed for the identification of five highly conserved sequences separated by regions of invariable length and diversified nucleotide composition (Zyskind et al. 1983). The conserved regions were used to define consensus sequence of R box (5′-TTATNCACA), which was correctly predicted as a binding site of DnaA (Zyskind et al. 1983). Biochemical studies confirmed subsequently that *oriC* contains five R boxes recognized by DnaA, three widely spaced high-affinity ones (R1, R2, and R4), and two additional, between them, which become occupied at higher concentrations of the initiator (R3 and R5 M) (Marguiles and Kaguni 1996; Schaper and Messer 1995). Detailed investigation of the DnaA interaction with *oriC* in the recent years proved that the intervening sequences, initially considered as spacers carrying little information, contain in fact a few types of low-affinity 9-mer sites, bound preferentially by DnaA–ATP (McGarry et al. 2004; Kawakami et al. 2005; Hansen et al. 2007). DAR has been shown to contain two oppositely oriented arrays of helically phased binding sites, one in each half of *oriC* (Rozgaja et al. 2011). In addition, doubts have been recently raised about the role of R3 (Rozgaja et al. 2011).

Only high-affinity R boxes are able to bind DnaA independently, in its both ATP- and ADP-associated forms, and interaction with other sequences requires cooperativity and DnaA–ATP (Speck et al. 1999; Messer et al. 2001; Miller et al. 2009). Members of nucleoid-associated proteins bind in the vicinity of each of the arrays, IHF in the left half and Fis in the right one (Gille et al. 1991; Roth et al. 1994). Both proteins bend DNA in the *origin* region (Gille et al. 1991; Roth et al. 1994), in the case of IHF, the introduced bend is very sharp (around 180 °C) and it most likely allows for interaction of the DnaA filament assembled at DAR with the DNA in the unwinding element (Ozaki and Katayama 2012).

For most of the duration of the cell cycle, the replication initiator protein remains bound to its high-affinity sites present at *oriC* (Miller et al. 2009). DnaA associated with R boxes marks *oriC* and serves as a landing pad for assembly of the preprimosomal complex, similar to the function of origin recognition complex (ORC) in eukaryotic cells. Under conditions supporting fast growth rates, Fis also remains bound to its recognition site within *oriC* for most of the cell cycle (Cassler et al. 1995). Fis blocks both IHF binding and association of DnaA with low-affinity sites at the left half of the origin (Ryan et al. 2004). Prior to initiation, the amount of DnaA–ATP molecules reaches a critical threshold, which results in the displacement of Fis and subsequent IHF-assisted occupation of the remaining sequences of lower affinity by the DnaA–ATP (Ryan et al. 2004). Thus, it was proposed that during the prereplication complex formation, DnaA oligomer formation initiates with extension from two high-affinity nucleation sites (R4 and R1) to the proximal weak site (Miller et al. 2009). Subsequently, filament builds up through progressive association of DnaA–ATP with the sites within the arrays. This process leads to the formation of DnaA complex at *oriC*, competent in unwinding of the AT-rich region and helicase loading (for a review, see Leonard and Grimwade 2011; Erzberger et al. 2006).

DnaA interacts with the single-stranded regions at DUE (Speck and Messer 2001; Duderstadt et al. 2010; Ozaki and Katayama 2011). Most likely, the first stage of filament formation involves interaction between DnaA protomers via domain I (extension from the high-affinity nucleation site to the proximal weak site), whereas filling in the arrays engages domain III-mediated interactions (Leonard and Grimwade 2011). The arrays of low-affinity sites do not span the entire length of the gaps between the high-affinity sites; therefore, it is not known whether DnaA molecules form a contiguous filament and the exact position of DnaA oligomers remains to be investigated (Leonard and Grimwade 2011). Binding of proteins within the gap regions and changes in the DNA structure, such as bending, evoked by binding of IHF may influence the filament geometry. Such staged assembly process implicates that many steps of oligomerization process may be affected by regulatory factors and that the particular configuration of these factors could vary with growth conditions. For instance, the cellular level of Fis is dependent on growth rate (Ball et al. 1992), which suggests that assembly of the DnaA oligomer might be differentiated according to growth conditions and Fis availability.

New levels of complexity have been recently added to this elegant model of assembly of prereplication complex. Detailed analysis of the structure of DnaA helical filament revealed that this oligomeric form is incompatible with biding of double-stranded DNA. Instead, domain IV is oriented toward the interior of the filament and docked against the AAA+ binding domain of the adjacent protomer (Duderstadt et al. 2010). Domain III elements form a pore along the DnaA filament, encompassing single-stranded DNA (Duderstadt et al. 2010). Since DnaA–ATP requires different conformations for association with double-stranded DNA and single-stranded DNA, initiation of DNA replication and *origin* melting involves a conformational change of the DnaA assembly (Duderstadt et al. 2010). Staged occupation of the high- and low-affinity binding sites within *oriC* by the DnaA would, thus, involve a conformation of DnaA with the DNA-binding domain (DBD) extending away from the body of the initiator to expose helix–turn–helix motif responsible for interaction with recognition sequences. In this form, interaction between the protomers of DnaA is less stable, but compensated by the close proximity of the weak binding sites. Origin unwinding would be performed by a different oligomeric state of DnaA, in which docking of DBD against AAA+ domain of the neighboring protomer would provide stability to the oligomer and allow for binding of ssDNA (Duderstadt et al. 2010). Moreover, it was demonstrated that in the presence of ATP, subunit/subunit interactions within DnaA oligomer facilitate melting of DNA strands by direct stretching the contacted strand (Duderstadt et al. 2011). In addition to this already complex picture of pre-RC formation, it was demonstrated that the left half of the *oriC* promotes the formation of DnaA oligomer competent in duplex opening and ssDNA binding, while the right half stimulates DnaB loading (Ozaki and Katayama 2012). These results were in agreement with earlier data showing that the right half of the origin is dispensable for slowly growing bacteria (Stepankiw et al. 2009). Differentiation in functions of the two parts of the *origin* is accompanied by a specific conformation acquired by the initiator protein bound to each of them (Ozaki et al. 2012). Although, most likely, the DnaA oligomer assembled at the entire *oriC* is required for cell cycle-regulated initiation of DNA replication, modularity of the initiator complexes formed at *oriC* provides additional possibilities of regulation of their activity.

How exactly the assembly of the DnaA oligomer competent for initiation of DNA replication is coordinated with cell growth and the division cycle remains still not fully understood. The prevailing view is that the initiation of DNA replication is linked to the growth-dependent accumulation of the initiator protein in its replication-proficient form (Donachie et al. 1968). In other words, DnaA–ATP accumulates to the amount sufficient for initiation when cells reach a particular mass. Recent results suggest that the total amount of DnaA, but not its concentration, is important for setting the initiation time (Hill et al. 2012). Interestingly, in

B. subtilis, contrary to the results obtained for *E. coli*, replication initiation control was not altered by the reduction of cellular mass (Hill et al. 2012).

The intracellular concentration of DnaA is constant in different media (Hansen et al. 1991), and thus, transcription of the *dnaA* gene was proposed to be dependent on the growth rate, in order to balance the differences in other cellular parameters at variable growth rates (Hansen et al. 1991; Chiaramello and Zyskind 1989). In fast growing cells, the synthesis of DnaA is more efficient than in slow growing bacteria, to ensure that replication can start at multiple *origins*. However, it is not known how DnaA accumulation is differentiated with respect to alterations in nutritional conditions. It has been proposed that changes in the level of the stringent response alarmone, ppGpp (Chiaramello and Zyskind 1990), produced in response to amino acid or carbon source limitation, may contribute to this regulation (for a review, see Potrykus and Cashel 2008). ppGpp was also proposed to play the main role in growth rate control of synthesis of the key macromolecules (Potrykus et al. 2011). Increase in intracellular levels of ppGpp blocks the initiation of DNA replication; however, the mechanism of its action with respect to DNA replication control remains poorly characterized. While down-regulation of the *dnaA* promoter activity may pertain to ppGpp-mediated replication arrest, it seems unlikely to be its only cause (Ferullo and Lovett 2008).

DnaA–ATP is also known to repress its own promoter (Speck et al. 1999). Based on this fact, it was proposed that autorepression of the *dnaA* promoter may depend on the growth rate. As a result, a constant amount of DnaA per unspecific binding sites present on the chromosome (DNA length), achieved due to the mentioned autoregulation of the *dnaA* promoter, would set the initiation timing. There is, however, no experimental data to support this model. It assumes that while affinity of DnaA to DnaA boxes present at the *dnaA* promoter varies with growth rate, possibly due to changes in DNA supercoiling, it does not show a similar dependence with respect to the binding sites present at *oriC* or that DnaA affinity changes for both sites in a identical way (Grant et al. 2011), and this aspect requires further investigation.

Besides regulation of DnaA synthesis, multiple mechanisms exist to ensure once-per-cell-cycle replication of chromosomal DNA. Control of the assembly of preinitiation complex by DnaA is achieved in *E. coli* cell essentially in two ways: by regulating its ability to oligomerize and by restricting its accessibility for binding at *oriC* (for a review, see Leonard and Grimwade 2010). As only DnaA–ATP is able to form the helical oligomer required for initiation, the first type of regulation is achieved mostly by influencing nucleotide-bound status of the initiator. The most important mechanism in this category is RIDA (regulatory inactivation of DnaA), which relies on stimulation of the ATP-ase activity of DnaA by the Hda protein, complexed with the β clamp of DNA polymerase (for a review, see Skarstad and Katayama 2013). This process is assumed to take place during the ongoing DNA synthesis and results in the formation of DnaA–ADP, inert for replication initiation.

Two other mechanisms ensure rejuvenation of the DnaA–ATP complex, namely its association with acidic phospholipids of the membrane and special DNA sequences present in the chromosome, called DARS (DnaA-reactivating sequence) (for reviews, see Leonard and Grimwade 2011; Saxena et al. 2013). Both these mechanisms stimulate exchange of ADP, bound to DnaA, for ATP. In addition, formation of the specific DnaA–ATP–*oriC* complex is promoted by the recently discovered DiaA protein (Keyamura et al. 2007).

The second type of control, preventing DnaA from binding to *oriC*, is accomplished in two ways. First, interaction of the initiator protein with the low-affinity sites within *oriC*, and thus premature reinitiation, is blocked immediately upon the replication start. This is done by the SeqA protein, associating with hemimethylated GATC sequences which overlap the lower-affinity DnaA boxes. In *E. coli* cells, Dam methylase modifies adenine at GATC sequences which become hemimethylated after synthesis of the new DNA strand. Occupation of hemimethylated GATC sites by SeqA at *oriC*, called sequestration, lasts for one-third of the cell cycle and concerns also the *dnaA* promoter region (for a review, see Waldminghaus and Skarstad 2009). Second, availability of DnaA is regulated through its titration by the *datA* region on the chromosome, containing DnaA-binding sites of unusually high capacity for DnaA, which were proposed to act as a reservoir of the protein in a manner dependent on nucleoid-associated protein—IHF (Nozaki et al. 2009). Recently, however, *datA* was also shown to stimulate ATP hydrolysis by the initiator protein, which is preceded by the formation of an oligomeric form of DnaA and probably requires specific inter-DnaA interactions, facilitated by IHF–mediated DNA looping (Kasho and Katayama 2013). Cell cycle analysis revealed that IHF binds to this region immediately after the initiation of DNA replication, at the time when RIDA is activated, and it was proposed that these two mechanisms act in concert to ensure coordination of DnaA inactivation within the cell cycle (Kasho and Katayama 2013). Furthermore, other DnaA boxes, abundant in the chromosome, compete with the *origin* for DnaA binding (for a review, see Skarstad and Katayama 2013), and due to the concurrent action of the above-mentioned mechanisms, dynamic distribution of DnaA molecules around the chromosome during the cell cycle likely contributes to timely provision of the correct amount of DnaA–ATP to *oriC* before the onset of replication (for detailed information, see Leonard and Grimwade 2011; Kasho and Katayama 2013). It is not known whether any of these regulatory mechanisms are subject to growth rate dependent control. Interestingly, it was demonstrated that a deletion of *ygfZ* gene suppresses overinitiation phenotype of *Δhda* mutants (Ote et al. 2006). The YgfZ protein was suggested to be a folate-binding protein involved in iron–sulfur cluster metabolism (Teplyakov et al. 2004; Hasnanin et al. 2012), and it was implicated in tRNA modification processes (Ote et al. 2006). YgfZ was proposed to repress DnaA–ATP hydrolysis or stimulate rejuvenation (Ote et al. 2006). What is more, it remains also uncharacterized whether assembly of the replication initiation complex and its composition are the same or differ under variable growth conditions.

In *B. subtilis*, no mechanisms inactivating replication potential of DnaA–ATP by converting it to DnaA–ADP are known. DnaA is regulated mostly by blocking its binding to *oriC* and oligomerization by the action of YabA, Soj, DnaN, and primosomal proteins DnaB and DnaD. YabA may provide the link between metabolic state of the cell and DNA replication control, since the YabA-dependent delay of DNA replication occurred under conditions supporting slow, but not fast, growth (Noirot-Gros et al. 2002; Hayashi et al. 2005).

Structural Organization of Cellular Processes and the Regulation of Cell Cycle

In this chapter, we would like to draw attention to macromolecular structure of the cell and its impact on the integration of processes and regulation of the cell cycle according to growth conditions.

Many years of biochemical and biophysical studies have accustomed us to think of proteins as highly purified entities that act in isolation, more or less freely diffusing until they find their cognate partner to bind to. While in vitro experiments largely remain the only way to investigate the intrinsic properties of molecules in detail, this approach ignores an important factor: In their natural milieu, proteins are surrounded by other molecules of different chemical nature, and this crowded environment can considerably modify their behavior. About 40 % of the cellular volume, on average, is occupied by all sorts of molecules. Furthermore, biological macromolecules live and operate in an extremely structured and complex environment within the cell, where they are subjected to crowding and confinement (Foffi et al. 2013).

E. coli cell consists of plethora of different compounds, while genetic material of this bacterium contains about 4,000 genes. The fundamental question to understand bacterial physiology, applying also to all living systems, is how cells manage to coordinate the usage of so many constituents to produce a phenotype adequate to external and internal conditions, yet retaining the ability to adapt to changing environment (Norris et al. 2013). Zooming into cellular processes, a similar question may be asked with respect to the regulation of bacterial cell cycle: How do cells integrate various external and internal signals to decide when to duplicate their genetic material and divide?

A hint to answer this question comes from large-scale proteomic studies performed in several bacterial species. Particularly, investigation of protein–protein interactions in one of the smallest bacteria, *Mycoplasma pneumoniae*, revealed a factory-like organization of the cell, where proteins are arranged into multi-subunit complexes, whereas around one-third of the uncovered interactions play a role of connectors, coupling various processes. These results suggested a higher level of proteome organization, involving extensive sharing of components between complexes and implying protein multifunctionality ("moonlighting") (Kuehner et al. 2009). In addition, many interactions inferred previously in proteome-wide studies of *E. coli* (Butland et al. 2005) were shown to be conserved in *M. pneumoniae* (Kuehner et al. 2005).

The picture of bacterial cell, where processes are orchestrated by the formation of macromolecular complexes and exchange of components between them, has been reflected in several recently published hypotheses on mechanisms governing cellular organization and physiology. The first of them is based on physicochemical properties of cytoplasm and suggests that in bacteria, subcellular structuring arises from localizations and interactions of biomacromolecules and from the growth and modifications of their surfaces by catalytic reactions (Spitzer 2011). Non-covalent interactions between macromolecules (including proteins) lead to the formation of large multicomponent molecular complexes (multiplexes) and local high degree of molecular crowding (supercrowding). Consequently, supercrowded complexes structure the cytosol into electrolyte pathways and nanopools that electrochemically "wire" the cell. This non-uniform crowding model allows for fast diffusion of biomacromolecules in the uncrowded cytosolic reservoirs, while the lower molecular weight metabolites are channeled due to the absence of sufficient free volume to attain bulk composition independent of their position. Most of the proteins were proposed to be multiplexed into assemblies associated with either cell envelope or the nucleoid and cross the less-crowded reservoirs by diffusion (Fig. 1a). Such organization would greatly reduce the infinite number of possible biochemical reactions occurring in unstructured cytoplasm (Spitzer et al. 2011).

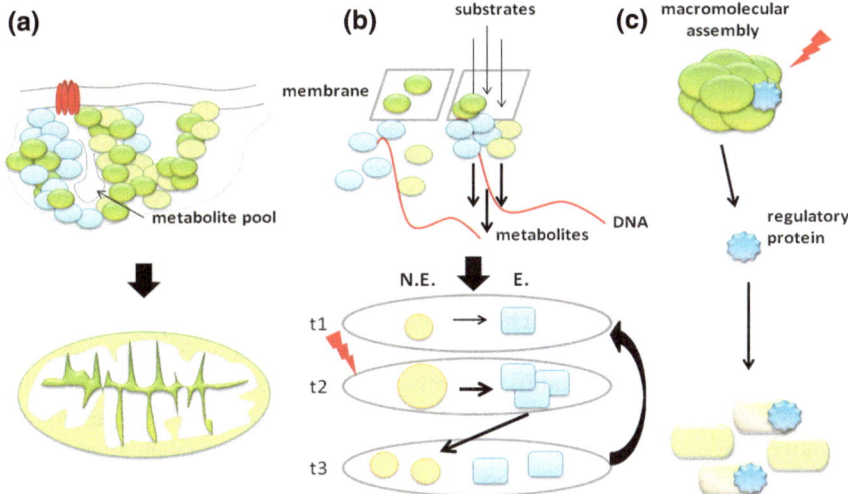

Fig. 1 Coordination of cellular processes at the level of macromolecular complexes. **a** Non-covalent interactions between biomolecules (*upper panel, spheres*) lead to high local degree of molecular crowding which structures the cytoplasm and channels metabolites. Multicomponent macromolecular complexes are proposed to be associated mainly with the cell membrane and the nucleoid (*lower panel, yellow* and *green compartments*), leaving the interior free for diffusion. **b** Hyperstructures represent an intermediate level of cellular organization between individual molecules and the cell as a whole. Non equilibrium hyperstructures may be composed of proteins, their cognate genes, and RNAs, and they are formed to perform a certain function and disassembled when no longer needed (*upper panel*). Cell cycle is driven by a balance between non-equilibrium and equilibrium hyperstructures (*lower panel*). In a growing bacterium (*gray ellipses*), the quantity of equilibrium hyperstructures (*blue rectangles*) relative to non-equilibrium ones (*yellow spheres*) increases. At a critical threshold, a signal is emitted (t2, *red lightning*), which results in the transition of some of the equilibrium material to non-equilibrium hyperstructures, and this triggers the initiation of DNA replication at the origins. **c** Macromolecular assemblies (*green spheres*), such as ribosomes and metabolons, may serve as depots of regulatory proteins (shown in *blue*), which are released upon a certain signal and control other cellular processes. The three hypotheses illuminate how reduction of compositional complexity of the cell is achieved to allow for coordination of different processes according to growth conditions (see text and references therein)

A related hypothesis proposes that macromolecular complexes serve as depots for regulatory proteins (Ray et al. 2013). Namely, multicomponent assemblies, such as ribosome, which perform essential cellular functions, were proposed to contain proteins that upon induced release acquire auxiliary functions and regulate other processes (Fig. 1c). The induction may occur in response to various external or internal signals, and upon freeing, the "daughter" protein may gain activity, after being inhibited in the parental complex, perform a similar function to that played in the parental complex in another cellular localization, or attain a different, "moonlighting" activity (Ray et al. 2013). This hypothesis provides means of communication between different processes and their coordination in accordance with current metabolic status of the cell. Further on, we will present examples pertaining to the regulation of bacterial cell cycle, where release of components from macromolecular assemblies (ribosome, metabolons) under certain conditions may contribute to the regulation of the DNA replication process.

Akin to the above-presented two hypothesis is the proposal that the cell cycle in bacterial cell is regulated by a balance between equilibrium and non-equilibrium hyperstructures, and in fact, it itself serves to maintain this balance, enabling the cell to retain the capability to grow at feast and to survive during famine (Norris and Amar 2012, and references therein). Non-equilibrium hyperstructures are transient macromolecular assemblies necessary for growth (composed of genes, mRNA, proteins, lipids, etc.), which are dependent on flow of energy and material (Fig. 1b). They are formed to perform certain functions and are disassembled when no longer required. On the other hand, equilibrium hyperstructures are essential for cell survival (for instance, they may be constituted by assemblies of enzymes in their inactive form). According to the hypothesis, maintaining intracellular balance between the two types of hyperstructures involves intensity and quantity sensing. The former relates to estimating the usage of non-equilibrium hyperstructures (for instance, the number of transcribing RNA polymerases per DNA unit) and is connected with nutrient availability and metabolic status, whereas the latter pertains to sensing the quantity of equilibrium structures by the cell. On attaining sufficient amount of equilibrium hyperstructures which may be inherited by the daughter cells and reaching by the non-equilibrium hyperstructures, the intensity which might limit further growth, signals (of various nature) would be emitted which would trigger initiation of the new cell cycle (Fig. 1b) (Norris and Amar 2012).

All the three hypotheses involve macromolecular complexes and communication between them as a higher level of cellular organization that enables integration of multiple internal and external signals to fewer outcomes which the cell can interpret. Several lines of evidence exist, showing that DNA replication is also carried out by macromolecular complexes or hyperstructures. Namely, SeqA protein, which binds hemimethylated DNA and is responsible for *origin* sequestration (for a review, see Waldminghaus and Skarstad 2009), was also shown to be involved in the organization of replication forks into conspicuous structures which can be microscopically visualized as SeqA foci (Molina and Skarstad 2004). The number of foci and the extent of their colocalization with the replication fork depend on the growth rate, and as no increase in their number is observed during replication initiation, new replication forks were proposed to be recruited to existing structures (Morigen et al. 2009; Fossum et al. 2007). This hyperstructure may also involve ribonucleotide reductase and other replication enzymes (Guzman et al. 2002, Molina and Skarstad 2004; Sanchez-Romero 2010).

Interestingly, in *B. subtilis*, replication proteins were shown to form a network of interactions, encompassing proteins engaged in distinct cellular functions: carbohydrate and amino acid metabolism, signal transduction, transcription, and chemotaxis (Noirot-Gros et al. 2002). Uncovering the rules regulating the formation of replication hyperstructures and their potential communication with other macromolecular assemblies engaged in essential cellular functions seems to be crucial for gaining insight into physiological regulation of DNA replication. Below, we describe recently demonstrated examples of functional and physical interactions between components of the replication machinery and other complexes, performing vital cellular processes, and discuss their role in the coordination of chromosomal DNA duplication with bacterial physiology.

The Link Between Metabolism and DNA Replication

The CCM is a set of biochemical pathways devoted to transport and oxidation of main carbon sources. In *E. coli*, it consists of phosphotransferase system, glycolysis, gluconeogenesis, the pentose–monophosphate bypass with the Entner–Doudoroff pathway, and Krebs cycle with the glyoxylate bypass and the respiratory chain. A direct link between metabolism and DNA replication has been suggested by a study exploring protein–protein interaction network associated with *Bacillus subtilis* replication machinery (Noirot-Gros et al. 2002). That work, exploiting yeast two-hybrid system, showed that several enzymes engaged in carbohydrate or amino acid metabolism interact in vivo with proteins involved in DNA replication. Importantly, an interaction was demonstrated between the DnaG primase and components of the large dehydrogenase complexes (PdhC, AcoC, BfmAB), catalyzing decarboxylation of acetoin and 2-oxo acids to generate acetyl-CoA and NADH. BfmAB was also found to associate with *B. subtilis* replicative helicase—DnaC (Noirot-Gros et al. 2002) (Table 1). Interestingly, PdhC—E2 subunit of pyruvate dehydrogenase complex—had been previously identified as a membrane-associated factor responsible for the inhibition of *B. subtilis* DNA replication (Stein and Firshein 2000). In addition, AcuB, another member of acetoin catabolism pathways, was singled out by yeast two-hybrid system as an interacting partner of YabA (Noirot-Gros et al. 2002) (Table 1), the regulator of helical filament formation by DnaA (Scholefield and Murray 2013). Thus, uncovering of a complex network of protein–protein interactions formed by the replisome provided an indication of a direct coordination of DNA replication with metabolic processes and other cellular activities; however, no functional relation between the respective proteins was characterized.

Table 1 Experimentally confirmed interactions of the replication machinery components with proteins engaged in transcription, translation, central carbon metabolism, and nucleoid dynamics

Replication protein	Interacting protein	Organism	Interaction outcome	References
DnaG	PdhC, E2 subunit of pyruvate dehydrogenase	*B. subtilis*	Unknown	Noirot-Gros et al. (2002)
DnaG	AcoC, acetone dehydrogenase, E2 component	*B. subtilis*	Unknown	Noirot-Gros et al. (2002)
DnaG	BfmBAB, 2-oxoisovalerate dehydrogenase (E1 beta subunit)	*B. subtilis*	Unknown	Noirot-Gros et al. (2002)
DnaC	BfmBAB, 2-oxoisovalerate dehydrogenase (E1 beta subunit)	*B. subtilis*	Unknown	Noirot-Gros et al. (2002)
YabA	AcuB, function unknown, (AcuA and AcuC, which belong to the same operon that takes part in acetate and fatty acid metabolism)	*B. subtilis*	Unknown	Noirot-Gros et al. (2002)
DnaA	RNA polymerase	*E. coli*	DnaA partially protects RNA polymerase from rifampicin inhibition	Flatten et al. (2009)
DnaA	L2, ribosomal protein	*E. coli*	L2 interacts with the N-terminal part of DnaA. Oligomerization of DnaA and helicase loading are inhibited	Chodavarapu et al. (2011)
DnaA	HU α subunit	*E. coli*	α subunit of HU interacts with the N-terminal region of DnaA. Interaction stabilizes DnaA oligomer at *oriC*	Chodavarapu et al. (2008a)
DnaA	Dps, nucleoid-associated protein	*E. coli*	Dps interacts with N-terminal region of DnaA and inhibits origin unwinding. Overexpression of Dps blocks initiation of DNA replication in synchronized cells	Chodavarapu et al. (2008b)

Another hint of the possible direct influence of metabolic pathways on the activity of replication proteins was obtained in genetic experiments carried out in *B. subtilis*. Temperature-sensitive phenotype resulting from changes in one of the proteins responsible for replication elongation DnaG (primase), DnaC (helicase; note that a homologous protein from *E. coli* is called DnaB), and DnaE (lagging strand DNA polymerase) was suppressed by mutations in genes encoding enzymes which carry out terminal reactions of glycolysis (*pgk*, *pgm*, *eno*, *pykA*) (Janniere et al. 2007). Since that work provided the first evidence for a direct genetic link between DNA replication and glycolysis, we consider that it deserves a more detailed description and discussion, which is presented below.

It is generally known that bacterial cells bearing mutations causing impairment of functions of replication elongation proteins at elevated temperatures (for mesophilic bacteria, such as *B. subtilis* or *E. coli*, it means usually 42–49 °C) form filaments and die shortly after a shift to restrictive temperature, due to inhibition or severe reduction of DNA synthesis (James 1975; Versalovic and Lupski 1997; Dervyn et al. 2001; Kawakami et al. 2001; Strauss et al. 2004). When looking for extragenic suppressors of temperature-sensitive mutations in the gene coding for *B. subtilis* DNA polymerase responsible for lagging strand DNA synthesis (*dnaE*), Janniere et al. (2007) mapped several suppressor mutations, allowing growth of the *dnaE* mutants at temperatures between 45 and 49 °C. To their surprise, the suppressor mutations were mapped to *pgk*, *pgm*, *eno*, and *pykA* genes, coding for enzymes catalyzing terminal reactions of glycolysis, rather than to genes which functions are related directly to DNA metabolism. Several experiments indicated that the suppression was direct. Namely, the parental temperature-sensitive *dnaE* mutants became temperature resistant after transferring the *pgkEP*, *pgm8*, *pgmIP*, or *pykAJP* mutations to them. Moreover, *dnaE* mutants carrying a suppressive mutation in *pgk*, *pgm*, *eno*, or *pykA* became temperature sensitive after introduction and expression of wild-type alleles of the glycolytic genes. In addition, artificial deletion of the *pykA* gene caused suppression of temperature sensitivity of the *dnaE* mutant.

The suppression of *dnaE* temperature sensitivity by mutations in *pgk*, *pgm*, *eno*, and *pykA* genes was particularly strong. The suppressed strains grew at restrictive temperatures for over 20 generations. They formed wild-type-like colonies and had a plating efficiency around 100 % under these conditions. Contrary to the temperature-sensitive *dnaE* mutants, suppressed strains did not form filaments after incubation at elevated temperatures. Moreover, the suppressed strains grew as fast as the corresponding metabolic mutants bearing the wild-type *dnaE* allele at restrictive temperatures (Janniere et al. 2007). Further studies demonstrated that such a strong suppression by mutations in *pgk*, *pgm*, *eno*, and *pykA* genes occurs also for temperature-sensitive mutants in *dnaC* and *dnaG* genes (Janniere et al. 2007). Therefore, replication functions of the lagging strand DNA polymerase (DnaE), DNA helicase (DnaC), and primase (DnaG) appeared to be somehow regulated by functions of enzymes involved in glycolysis.

In a series of elegant experiments, Janniere et al. (2007) indicated that the suppression of mutations in replication genes by specific mutations in genes encoding glycolytic enzymes is direct rather than indirect. They found that the suppression of temperature sensitivity of *dnaE, dnaC,* and *dnaG* mutants could not be caused by osmotic, energetic, or nutritional stresses, i.e., under conditions which one might suppose to occur in cells devoid of *pgk*, *pgm*, *eno*, and *pykA* functions. Moreover, the suppression did not depend on the growth rate decrease. Finally, they have demonstrated that the suppression did not depend on accumulation of the temperature-sensitive replication proteins, but rather these proteins, products of the mutated *dnaE, dnaC,* and *dnaG* alleles, became temperature resistant in cells bearing the suppressor mutations.

The question remained what is a mechanism for a direct link between regulation of DNA replication and terminal reactions of glycolysis. Janniere et al. (2007) suggested that activity of this part of glycolytic pathway evokes conformational changes in the replisome components. Such alterations might result from protein binding to metabolites, post-translational modifications, or physical protein–protein interactions. Moreover, Janniere et al. (2007) suggested that the link between DNA replication and cell metabolism (particularly CCM) may be ubiquitous not only among bacteria but also in eukaryotic organisms. If so, determining the mechanisms of this link may be important to understand more complex processes, such as cell cycle control and carcinogenesis. Early events of the latter process include stimulation of glycolysis (the Warburg effect) and a decrease in DNA replication fidelity (Loeb et al. 2003; Gatenby and Gillies 2004). Therefore, one can speculate that perturbations of the link between DNA replication and metabolism might contribute significantly to carcinogenesis.

The question whether the replication–metabolism link is ubiquitous (at least among bacteria) or unique to *B. subtilis* has been addressed subsequently. In fact, the interplay between CCM and DNA replication was demonstrated to occur in *E. coli* (Maciag et al. (2011), corroborating the proposal by Janniere et al. (2007) about universality of this phenomenon. In *E. coli*, effects of mutations in *dnaE* (encoding the α subunit of DNA polymerase III), *dnaN* (coding for DNA polymerase III β clamp), and *dnaG* (coding for the primase) were partially suppressed by deletions of genes coding for enzymes involved in glycolytic, acetate overflow, and pentose–phosphate pathways (Maciag et al. 2011) (Fig. 2). Interestingly, the strongest suppression was observed in the case of a mutation in the gene coding for the replication initiator—*dnaA*. Defects of the DnaA46 protein, resulting in temperature-sensitive growth of strains bearing the mutant initiator, were overcome by the absence of enzymes comprising acetate overflow mechanism (*pta* and *ackA*) (Fig. 2). As in the case of the study on *B. subtilis* (Janniere et al. 2007), the suppression of temperature sensitivity of replication mutants by dysfunctions of genes coding for CCM enzymes was direct rather than indirect. First, expression of appropriate wild-type allele of the CCM gene reversed effects of temperature sensitivity suppression by the corresponding mutant allele. Second, although in most cases, the growth rates of the double mutants revealing suppression of the temperature sensitivity were lower at 30 °C than in wild-type bacteria, a similar or lower decrease in the growth rate was observed also in double mutants which did not suppress the temperature sensitivity; thus, the observed suppression effects could not be caused simply by a decrease in bacterial growth rate. The latter conclusion has been supported by results of experiments in which the suppression could not be achieved by the growth of the replication mutants in media containing various carbon sources which allow for different growth rates (Maciag et al. 2011).

The results reported by Maciag et al. (2011) and summarized above indicated that at least in *E. coli*, direct metabolic control may exist not only at the elongation stage of DNA replication but also at its initiation phase. This is also interesting in

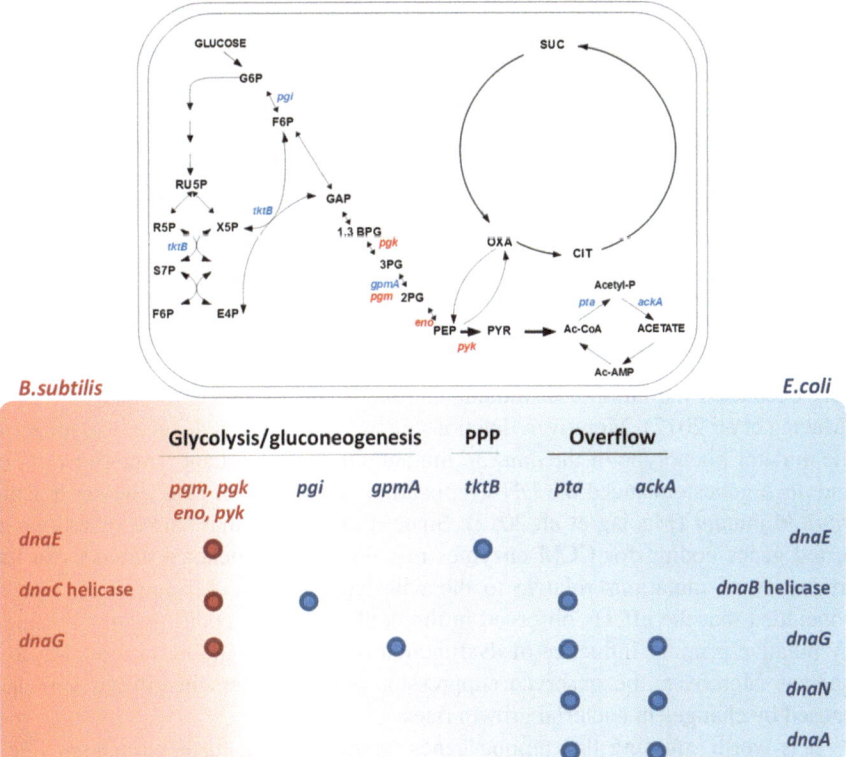

Fig. 2 The scheme of the central carbon metabolism (CCM, *upper panel*), with indicated genes coding for enzymes involved in particular reactions. Lower panel demonstrates the pattern of suppressions of effects of mutations in genes coding for replication factors in *B. subtilis* and *E. coli* by particular mutations in genes coding for CCM enzymes, involved in glycolysis/gluconeogenesis, pentose phosphate pathway (*PPP*), or the overflow reactions. In both panels, *red color* indicates specific suppressions in *B. subtilis*, and *blue color* indicates specific suppressions in *E. coli*. Abbreviations: **1,3-BGP** 1,3-biphosphoglycerate, **2PG** 2-phosphoglycerate, **3PG** 3-phosphoglycerate, **G3P** galactose-3-phosphate, **G6P** glucose-6-phosphate, **F6P** fructose-6-phosphate, **OXA** oxaloacetate, **FBP** fructose-1,6-bisphosphate, **GAP** glyceraldehyde-3-phosphate, **PEP** phosphoenolpyruvate, **PYR** pyruvate, **Ru5P** ribulose-5-phosphate, **R5P** ribose-5-phosphate, **S7P** sedoheptulose-7-phosphate, **E4P** erythrose-4-phosphate, **Ac-CoA** acetyl coenzyme A, **Acetyl-P** acetyl phosphate, **Ac-AMP** acetyl-AMP, **CIT** citrate, **SUC** succinate, **X5P** xylulose-5-phosphate. From Barańska et al. (2013)

light of recent findings showing that timing of the DNA replication initiation is strictly coupled to cell size (mass) in *E. coli* but not in *B. subtilis* (Hill et al. 2012), emphasizing that the specific mechanisms of coupling DNA replication to cell growth differ among bacteria. However, general solutions ensuring this coordination, including a direct link between CCM and DNA replication, are preserved. Like in studies on *B. subtilis* (Janniere et al. 2007), experiments performed on *E. coli* demonstrated that most (but not all) of the genetic changes that restored

viability of replication mutant cells at increased temperatures alleviated also fila-
mentation and aberrant chromosome positioning caused by the defects of the replication machinery (Maciag-Dorszynska et al. 2012).

Intriguingly, alterations in carbon metabolism enhanced or suppressed also
mutator phenotypes of *dnaQ49* and *dnaX36* mutant strains (Maciag et al. 2012).
The *dnaQ* gene codes for the ε subunit of DNA polymerase III, and *dnaQ49* is
a recessive allele which confers a temperature-sensitive phenotype of decreased
fidelity of DNA synthesis (Nowosielska et al. 2004). The *dnaX36* mutation causes
a dysfunction of the τ subunit DNA polymerase III, which results also in a mutator effect (Gawel et al. 2011). In two independent assays measuring frequency of
appearance of spontaneous (i.e., not induced by chemical of physical mutagens)
mutations, it was found that deletions of *pta, ackA, acnB,* and *icdA* genes considerably decreased the number of mutants among bacteria bearing the *dnaQ49* allele
(Maciag et al. 2012). Moreover, deletions of *zwf, acnB,* and *icdC* genes suppressed
the mutator phenotype of the *dnaX36* mutant. On the other hand, mutations in *pta*
and *ackA* genes enhanced the DNA replication fidelity defect characteristic for the
dnaX36 mutant (Maciag et al. 2012). Since it was found that single mutations in
tested genes coding for CCM enzymes revealed no significant differences in the
frequency of mutations relative to the wild-type bacteria, Maciag et al. (2012)
concluded that the effects observed in the double mutants could not be explained
by putative primary influence of dysfunction of the CCM genes on mutation frequency. Moreover, the observed suppression and enhancement effects were not
caused by changes in bacterial growth rates.

It is worth stressing that among genes whose deletion further impaired fidelity of replication in the above-mentioned mutators were *pta* and *ackA*, which
encode enzymes involved in the acetate/acetyl-CoA pathway (Maciag et al. 2012).
Changes in the same metabolic genes were also identified to cause synthetic inhibition of growth in combination with defects in DNA recombination/repair genes
recA and *recBC* (Shi et al. 2005). Comparison of the effect of *pka/ackA* dysfunction on the ability of growth of other recombination mutants resulted in proposal
that the inhibition of colony formation by *ptk/ackA rec* double mutants results
from accumulation of double-strand breaks caused by destruction of acetate overflow mechanism (Shi et al. 2005). This, in turn, might suggest impairment of replication machinery in such strains.

Another example of interrelation between carbon metabolism and DNA replication has been uncovered recently. Namely, it was found that an absence of
AspC function leads to generation of small cells, decreased frequency of initiation of DNA replication, and slower growth (Liu et al. 2014). The *aspC* gene
encodes an aminotransferase involved in synthesis of aspartate from oxaloacetate.
Overproduction of AspC in the *ΔaspC* background leads to an increase in the
number of origins per cell, and similar effect was evoked by addition of aspartate to a wild-type strain culture medium. In addition, the presence of aspartate
was accompanied by higher growth rate and enlarged cell size. Interestingly, of 20
amino acids, only aspartate affected both DNA replication and cell size, whereas
glutamate stimulated the number of origins per cell and growth rate. The absence

of AspC resulted in a decrease in the amount of the DnaA protein per cell, but did not change its concentration, suggesting that aspartate metabolism may indirectly regulate initiation of DNA replication by affecting DnaA availability for *oriC*. In *E. coli*, amino acid and sugar metabolisms are interlinked, and aspartate serves also as a precursor for the synthesis of other amino acids, pantothenic acid (precursor of CoA), NAD, and nucleotides. Therefore, it was suggested that aspartate metabolism may serve as a hub, interlinking carbon metabolism with DNA replication, cell growth, and division (Liu et al. 2014). It is worth noting that a direct link between metabolism and cell size control has been recently established in three model bacterial organisms. It was demonstrated that in *B. subtilis* and *E. coli*, UDP–glucose signals nutritional status to the cell division machinery *via* an interaction with functional homologues, UgtP and OpgH glycosyl transferases, respectively. Under conditions supporting fast growth rates, UDP–glucose accumulates and glycosyl transferases delay cell division by directly inhibiting FtsZ assembly until the cell reaches a proper size (Weart et al. 2007; Hill et al. 2013). Conversely, low levels of UDP–glucose during slow growth result in division occurring at lower cell length. Also, in *C. crescentus*, KidO—a NAD(P)H oxidoreductase homolog—inhibits FtsZ ring formation, and therefore, it was suggested that cell size is coordinated in this bacterium by NAD(P)H level. It was proposed that in *E. coli*, aspartate metabolism may affect cell size by modulating UDP–glucose concentration, coordinating timing of the initiation of DNA replication with cell division (Liu et al. 2014).

Here, we would like to come back to the suggestion that if the replication–CCM link is ubiquitous and occurs in most, if not all, organisms, it may have serious implications for not only regulation of DNA replication, but also more complex processes, including carcinogenesis (Janniere et al. 2007; Barańska et al. 2013). About forty years ago, the mutator hypothesis was proposed (Loeb et al. 1974) to explain that a high number of mutations are usually needed for carcinogenesis, while there is a relatively low level of spontaneous mutation rate in normal cells. Subsequent works confirmed that multiple genetic changes are required to cause malignancy and that tumors exhibit different types of genetic instability. More recent studies indicated that apart from chromosomal aberrations and microsatellite instability, well documented in the literature, base substitutions and small deletions/insertions play important roles in carcinogenesis (reviewed and summarized by Preston et al. 2010). This point mutation instability is an important pathway to cancer as single nucleotide sequence changes can activate oncogenes and inactivate tumor suppressors. Therefore, if the link between CCM and DNA replication fidelity, demonstrated in *E. coli* by Maciag et al. (2012), exists also in human cells, one might suppose that specific changes in efficiencies of particular CCM reactions and/or metabolic variations connected to CCM could significantly affect the mutation rates during human DNA replication and thus considerably influence the carcinogenesis.

Although the nature of the interplay between components of the replication complex and metabolic pathways is currently unknown, there are several possibilities how changes in CCM enzymes may influence DNA replication process. Firstly, some of the intermediate metabolites serve as signaling molecules, and their accumulation may lead to global changes of cellular physiology. For instance, a number of studies suggested that acetyl phosphate, produced by phosphotransacetylase (Pta) during acetogenesis or by acetate kinase (AckA) during acetate degradation, most likely plays a role for a phosphoryl donor to a subset of two-component response regulators that control diverse cell functions (reviewed by Wolfe 2010). Secondly, imbalanced level of metabolites, for example acetyl-CoA, or lack of the activity of some enzymes may directly cause changes in post-translational modifications and hence activity of the replication apparatus (for instance, acetylation, in the absence of Pta-AckA pathway). Phosphorylation of single-stranded DNA (ssDNA)-binding (Ssb) proteins was demonstrated to occur in *B. subtilis* and *E. coli* and to modify their affinity for ssDNA (Mijakovic et al. 2006).

Another possibility is that protein–protein interactions between metabolic enzymes and components of the replication machinery are absent or altered in the studied mutants. Although no physical association was proven in the cases described above, the properties of some of the glycolytic enzymes described in both prokaryotic and eukaryotic organisms imply such a possibility. It was shown that several of these metabolic proteins (hexokinase, lactate dehydrogenase, enolase, glyceraldehyde-3-phosphate dehydrogenase) are multifunctional ("moonlighting") enzymes, and except their role in carbohydrate utilization pathways, they take part in the regulation of distinct processes such as transcription, apoptosis, and motility (reviewed in Sirover 2011; Kim and Dang 2005).

Of particular interest are functions of glyceraldehyde-3-phosphate dehydrogenase (GAPDH) and lactate dehydrogenase (LDH) in the activation of the histone H2B promoter during the S phase in mammalian cells (reviewed by He et al. 2012). In this case, GAPDH is a key component of the H2B promoter activating complex (OCA-S) (Zheng et al. 2003). GAPDH activates transcription using NAD(H) as a cofactor (Zheng et al. 2003), whereas LDH—when present in OCA-S—converts NADH to NAD^+ (Dai et al. 2008). Since transcription of the gene coding for histone H2B is confined to a certain NAD^+/NADH ratio (Dai et al. 2008), GAPDH and LDH most likely control transcription and S phase progression via sensing intracellular redox status (Yu et al. 2009). It was also proposed that OCA-S-mediated regulation, modulated via NAD(H), may be directly applied also to DNA-synthesizing machinery (He et al. 2012). Thus, in eukaryotic cells, moonlighting activities of CCM enzymes coordinate transcription and replication of the genetic material in accordance with metabolic status of the cell. It is possible that similar mechanisms operate also in bacteria. Recently, GAPDH of *E. coli* was shown to interact in vivo with a bunch of proteins engaged in various processes, including DNA repair (Ferreira et al. 2012).

Transcriptional Activation: The Unsolved Problem

In early studies on DNA replication in *E. coli*, it was noticed that rifampicin, which inhibits RNA polymerase activity, prevents also initiation of DNA replication both in vivo and in a reaction performed in crude extracts. This led to the proposal of a direct engagement of RNA polymerase and transcription process in the activation of *oriC* for replication. This conclusion was further supported by the observation that thermal sensitivity of certain *dnaA* mutants can be suppressed by a secondary mutation in the *rpoB* gene, encoding the β subunit of RNA polymerase (Atlung 1984). Allele specificity of this suppression was suggested to be an indication for a direct interaction between RNA polymerase and the DnaA protein. Importantly, formation of a physical complex between RNA polymerase and DnaA has recently been confirmed in vitro (Flatten et al. 2009). However, physiological importance of this interaction has not been clarified. DnaA regulates also transcription from several promoters, including the promoter of the *dnaA* gene (reviewed in Messer and Weigel 2003), and thus, it is not clear whether the interaction between RNA polymerase and the replication initiator plays a role in the regulation of transcription or control of DNA replication, or both.

Efforts to identify promoters responsible for the transcriptional activation step have been unsuccessful. The best candidates for the regulation of *oriC* activity seemed to be two promoters flanking this region: *mioC* and *gidA*, whose activities were found to fluctuate in accordance with the cell cycle. Transcription from *mioC*, situated downstream of *oriC* and directing its transcription toward DAR, is shut down prior to initiation, whereas activity of *gidA,* residing upstream of the AT-rich region, transcribing away from *oriC,* is repressed just after the start of replication (Theisen et al. 1993). Deletion of these promoters, however, had no effect on the initiation of chromosomal DNA replication, although minichromosomes devoid of any of them displayed perturbations in replication and stability (Bates et al. 1997; Lobner-Olesen and Boye 1992). Thus, transcriptions from *gidA* and *mioC* promoters do not regulate replication from *oriC*; however, their putative effects on the topology of this DNA region may play a role under unfavorable conditions. In fact, activity of *gidA* was demonstrated to be necessary in the absence of DnaA box R4 (Bates et al. 1997). Furthermore, a constitutive expression from *mioC* was shown to increase thermal sensitivity of some *dnaA* mutants and suppress the phenotype of *dnaAcos*, a mutant which at 30 °C reveals overinitiation of DNA replication and subsequent growth arrest (Su'etsugu et al. 2003).

Recently, new light was shed on the potential role of transcription and associated changes in DNA topology on the regulation of the *oriC* activity. In a series of elegant experiments, Kaur and colleagues have shown that none of the high-affinity DnaA-binding sites in *oriC* is essential for its function however, loss of two such sites inactivates the *origin*. Moreover, *oriC* lacking R1 or R4 became dependent on the function of IHF and Fis. In addition, binding DnaA to high-affinity sites abrogated spontaneous unwinding of the 13-mere region, which was not observed in the case of *origin*s lacking a functional high-affinity binding site. Elimination of R4 abolished also Fis-mediated repression of DnaA and IHF binding to the left half of *oriC*. On the basis of these results, a model was proposed in which high-affinity sites have a role in establishing a special conformation of *oriC* that restricts the amount of bending tolerated by *oriC* and allows for switch-like transition from ORC to pre-RC. DNA topology, similarly to the effect observed in the case of LacR or λ CI-mediated repression of transcription (Czapla et al. 2013;

Norregaard et al. 2013), may influence loop stability and DnaA protein ability to form the effective *oriC* conformation. Dependence of *oriC* function on the *gidA* promoter activity in the absence of R4 DnaA-binding site may support this notion (Bates et al. 1997). This model would also account for the role of HU protein, which could provide necessary degree of flexibility to the formed isolated DNA domain (Becker et al. 2005).

Although the mechanism(s) of regulation of *oriC* activity by transcription still remains elusive, a hint for its possible role comes from the studies on DNA replication control of bacteriophage λ. The *origin* of bacteriophage λ DNA replication, called *ori*λ, is located in the middle of the *O* gene (reviewed in Wegrzyn and Wegrzyn 2005; Wegrzyn et al. 2012). The *O* gene codes for the replication initiator protein, which binds to the replication *origin*, forming the nucleoprotein structure called "O-some." The second λ replication protein, the *P* gene product, is involved in delivery of the host (*E. coli*)-encoded DNA helicase, the DnaB protein, to the O-some. The formed *ori*–O–P–DnaB structure, called "preprimosome," is stable but inactive in promoting DNA replication due to strong interactions between P and DnaB proteins which prevent the helicase activity of the latter component. Therefore, remodeling of the preprimosome is necessary, which is performed by the action of heat-shock proteins (molecular chaperones): DnaK, DnaJ, and GrpE. Importantly, heat-shock-protein-dependent preprimosome remodeling is coupled with transcriptional activation of *ori*λ, a process of transcription proceeding through and further downstream of the replication *origin*. This transcription process is necessary for efficient initiation of λ DNA replication in vivo even if all replication proteins are provided (reviewed in Wegrzyn and Wegrzyn 2005; Wegrzyn et al. 2012). Studies demonstrating that λO replication initiator enhances transcription-induced supercoiling by DNA gyrase and has an ability to form topologically isolated domain suggested the mechanism of transcriptional activation based on the changes in DNA topology, introduced by RNA polymerase (Leng and McMacen 2002; Leng et al. 2011). Results of recent studies indicated that RNA polymerase directly interacts with the O replication initiator protein, which, contrary to DnaA, is not involved in transcriptional regulation (Szambowska et al. 2011). Thus, the formation of a complex between RNA polymerase and the replication initiator may influence prereplication complex assembly and affect transcription-coupled changes of DNA topology. Notably, in the case of bacteriophage λ, transcription was shown to regulate directionality of replication starting from *ori*λ (Baranska et al. 2001). For *E. coli*, it was also suggested that the helicase moving leftward from the origin is loaded first into the replication forks, to maintain stability of the initiator complex before the formation of the rightward replisome (Breier et al. 2005), whereas DNA topology was shown to influence the timing of the release of replication forks.

The above-listed findings make the transcription-coupled topological changes of DNA a likely candidate to regulate the replication initiation step (Smelkova and Marians 2001). It was demonstrated that rifampicin causes global chromosome decompaction, which could be the cause of replication initiation arrest (Cabrera et al. 2009). Similar effect is also exerted by guanosine tetraphosphate (ppGpp), an alarmone of the stringent response (the bacterial response to nutritional deprivation, particularly amino acid starvation; for a review, see Potrykus and Cashel 2008), due to inhibition of transcription from rRNA promoters (Cabrera et al. 2009). The mechanism of the inhibitory effect of ppGpp on the initiation of DNA replication in *E. coli* also remains obscure. Intriguingly, ppGpp-mediated arrest of the initiation of DNA replication is abolished in *seqA* mutants, and this effect is largely independent of the sequestration of *oriC* (Ferullo and Lovett 2008). It was demonstrated that *seqA* mutants have altered the level of chromosomal DNA supercoiling and displayed global changes of transcription (Klungsoyr and Skarstad 2004; Lobner-Olesen et al. 2003). Thus, both global chromosome structure and more local changes of DNA topology, driven by transcription, may play a role in coupling of physiological state of the cell to DNA replication.

The above-mentioned stringent control alarmone, ppGpp, is one of the factors linking bacterial cell metabolism to regulation of RNA synthesis and thus indirectly but significantly influencing the control of bacterial DNA replication. Although the mechanisms of ppGpp-mediated regulation of replication of *E. coli* chromosome are still not fully elucidated, it appears clearly that another replicon functioning in this bacterium, λ plasmid, is negatively controlled by this nucleotide due to inhibition of activity of the p_R promoter (Wróbel et al. 1998) and resultant inefficient transcriptional activation of the *origin* (for a review, see Węgrzyn and Węgrzyn 2005). Indeed, replacement of p_R with another promoter, insensitive to ppGpp, abolished the stringent response-dependent negative regulation of λ plasmid replication initiation (Szalewska-Pałasz et al. 1994).

In *E. coli*, ppGpp binds to RNA polymerase, directly influencing the regulation of transcription initiation from vast majority of promoters (Ross et al. 2013; Zuo et al. 2013). However, it was also demonstrated that DnaG primase is another direct target for ppGpp action in both *E. coli* and *B. subtilis*; this nucleotide inhibits enzymatic activity of the DnaG protein (Wang et al. 2007; Maciąg et al. 2010; Rymer et al. 2012). Despite similar responses of DnaG primase to ppGpp in vitro, earlier studies demonstrated that the stringent response alarmone inhibits DNA replication at the initiation step in *E. coli* but at the elongation phase in *B. subtilis* (Levine et al. 1991). Although more recent and more detailed studies have shown that DNA replication elongation rate is decreased by high levels of ppGpp also in *E. coli*, this effect was far less pronounced than in *B. subtilis* (DeNapoli et al. 2013). Perhaps surprisingly, evidence was demonstrated that while ppGpp influences the DNA replication elongation in *E. coli* cells only weakly, it strongly inhibits DNA replication reconstructed in vitro from *E. coli* proteins (Maciąg-Dorszyńska et al. 2013). To explain this ostensible paradox, a hypothesis was proposed, suggesting that although ppGpp inhibits activities of DnaG primases from *E. coli* and *B. subtilis* to similar extent in vitro, this nucleotide might be unable to efficiently block DNA replication elongation in *E. coli* cells due to its strong interactions with abundant RNA polymerase molecules (Maciąg-Dorszyńska et al. 2013). Thus, RNA polymerase would outcompete DnaG primase from ppGpp binding in this bacterium, resulting in only minor effects of the stringent response factor on DNA replication elongation (Barańska et al. 2013). There is a complete different scenario in *B. subtilis* cells. RNA polymerase of this bacterium does not bind ppGpp, and transcription inhibition during starvation is caused by changes in nucleotide pools (Krasny and Gourse 2004; Toyo et al. 2008). Therefore, the *B. subtilis* DnaG primase is not outcompeted by RNA polymerase for interactions with ppGpp and can be inhibited not only in vitro but also inside the cells. This can result in the inhibition of DNA replication elongation (Barańska et al. 2013; Maciąg-Dorszyńska et al. 2013).

It is worth emphasizing that growing understanding of bacterial chromosome organization and dynamics suggests that the chromosome itself is one of the key elements orchestrating cellular processes in accordance with environmental conditions and defining cellular architecture (reviewed in Muskhelishvili and Travers 2013; Dorman 2013; Ptacin and Shapiro 2013; Wang et al. 2013). The chromosome of *E. coli* is compacted more than 1000-fold to fit inside the bacterial cell, and the principal mechanism by which this compaction is achieved is negative supercoiling. The level of chromosomal supercoiling regulates transcriptional activity of many promoters (Peter et al. 2004; Blot et al. 2006), and in turn, it reflects the metabolic status of the cell (van Workum et al. 1996). It has been demonstrated that DNA superhelicity influences DNA-binding properties of many proteins (reviewed in Fogg et al. 2012); however, the exact impact of DNA topology on the formation of the replication initiation complex remains poorly investigated. Even less is known about its effect on association of DnaA with other chromosomal regions. Importantly, detailed studies on chromosome structure in *E. coli* and *C. crescentus* have proven that transcription is a major structuring force (Cagliero et al. 2013; Le et al. 2013). The above-mentioned investigation in *C. crescentus* and earlier genetic studies on *Salmonella enterica* demonstrated that transcription of highly expressed genes forms boundaries of supercoiled chromosomal domains (Le et al. 2013; Booker et al. 2010; Deng et al. 2004). In *E. coli*, genes that are highly active during exponential phase and down-regulated during induction of the stringent response displayed a high level of spatial clustering (Cagliero et al. 2013, Jin et al. 2013). Remarkably, position of a gene on the chromosome and overall nucleoid compaction has been shown to determine LacI protein distribution within the cell (Kuhlman and Cox 2012). Thus, activity of RNA polymerase has a great impact on chromosome structure (and vice versa), and this way, it may also influence formation and localization of protein complexes, potentially also those controlling DNA replication.

A striking example of mechanisms correlating metabolism, DNA structure, transcription, and DNA replication comes from recent studies performed on yeast. Continuously grown cultures of yeast can become spontaneously self-synchronized. During ultradian cycle, cells oscillate between phases of increased and decreased respiration (oxidative and reductive phases, respectively), which results in concomitant alterations in ATP:ADP ratio (Klevecz et al. 2004; Machne and Murray 2012). Oscillations are reflected in the whole transcriptome, organizing it into two superclusters of genes related to cellular growth and anabolism expressed during the phase of high oxygen uptake, and catabolism and stress response genes, active during low oxygen uptake. The mechanism of the differential regulation of the two clusters involves differences in histone occupancy between their promoters and differential effect of ATP-dependent chromatin remodeling machineries associated with each cluster (Machne and Murray 2012). Importantly, DNA-replicating cells are enriched during the reductive phase of each cycle, which may constitute a mechanism protecting DNA replication from the oxidative damage (Klevecz et al. 2004). It is not known whether bacteria undergo similar oscillations between anabolic and catabolic activities, but it was proposed that concomitant alterations in gyrase activity could lead to changes in DNA supercoiling around effectively transcribed genes and their subsequent spreading across the genome (Rovinskiy et al. 2012).

Activity of DnaA Is Inhibited by Components of Translation Machinery

In bacteria, the number of ribosomes in the cell rises proportionally to growth rate, in accordance with increasing demands for protein synthesis (Bremer and Dennis 1996; Keener and Nomura 1996). Although the level of ribosomal proteins (RP) in *E. coli* is coordinated with rRNA synthesis via translational repression of mRNA by RPs (reviewed in Nomura et al. 1999), a substantial amount of free ribosomal proteins is present in the cytoplasm (Ulbrich and Nierkhaus 1975), and they also become released from the ribosome upon stress (Zundel et al. 2009). Many RPs, especially in eukaryotic organism, have been demonstrated to perform extraribosomal functions (reviewed by Warner and McIntosh 2009). In *E. coli*, the L4 protein was shown to modulate activity of RNase E and thus to influence degradation of many mRNAs (Singh et al. 2009). Such "moonlighting" activities of RPs may coordinate distinct cellular functions with growth or in response to adverse conditions. Intriguingly, a component of the ribosome was also shown to affect activity of replication proteins in *E. coli*. Formation of a complex between ribosomal protein L2 and DnaA replication initiator was first suggested by results of proteomic studies (Butland et al. 2005). Subsequently, L2 was reported to interact directly with the N-terminal domain of DnaA (Chodavarpau et al. 2011). This interaction destabilized an oligomer formed by DnaA at *oriC* and hence impeded DnaB loading (Fig. 2). Two forms of the L2 ribosomal protein, full length and C-terminally truncated (dubbed tL2), were associated with the replication initiator with essentially the same outcome. Overexpression of either L2 or tL2 reduced colony formation by wild-type strain, and this effect was exacerbated in *recA* and *recB* recombination mutants. However, excess of tL2 failed to block initiation of replication in *dnaC*ts synchronized cells (Chodavarpau et al. 2011). This might suggest that special conditions must be met to evoke L2-mediated inhibition of DnaA activity in vivo.

Like in bacteria, cell growth is a prerequisite for cell proliferation also in eukaryotes. In response to proliferation stimuli, protein synthesis is enhanced, and consequently, the increased anabolic demand is accomplished by elevated rate of ribosome biogenesis (reviewed by Lempiainen and Shore 2009). Interestingly, it is now well documented that in vertebrate cells, the surveillance of ribosome assembly is a mechanism coupling cell growth to cell cycle, and aberrations in ribosome biosynthesis led to the inhibition of cell cycle progression (reviewed in Deisenroth and Zhang 2010 and 2011). This control is performed by several ribosomal proteins, which stabilize p53 tumor suppressor protein, resulting in the arrest of the G_1–S phase transition. Recently, cell cycle was showed to be controlled in response to alterations in ribosome biogenesis also in *Drosophila*, in yeast, and in a p53-independent manner in mammalian cells, suggesting that this might be a ubiquitous and conserved mechanism (Donati et al. 2012). Furthermore, in *Saccharomyces cerevisiae*, the Yph1p protein, engaged in ribosome biogenesis, interacts with the origin recognition complex (ORC), and its depletion leads to the arrest in G_1 or G_2 phase (Du & Stillman 2002). Thus, it is possible that bacteria employ similar tools to regulate DNA replication in accordance with the status of translation machinery.

Nucleoid-Associated Proteins Modulate Activity of the DnaA Initiator

Nucleoid-associated proteins play a vital role in the organization of bacterial nucleoid. Owing to their capabilities to influence DNA topology and stabilize different DNA structures, they regulate DNA transactions: transcription, recombination, and DNA replication (for a review, see Browning et al. 2010). At *oriC*, IHF and Fis were shown to govern prereplication complex assembly and origin unwinding by influencing DNA structure (IHF) and physically blocking DnaA binding to a subset of lower-affinity binding sites (Fis) (for a review, see Leonard and Griwade 2011; Ozaki and Katayama 2011).

Interestingly, two nucleoid-associated proteins were shown to interact physically with DnaA and affect its ability to initiate DNA replication. Namely, HU, a dimeric protein, encoded by two closely related genes, is expressed in a growth phase-dependent manner (Claret and Rouviere-Yaniv 1997) (Fig. 3). It binds to

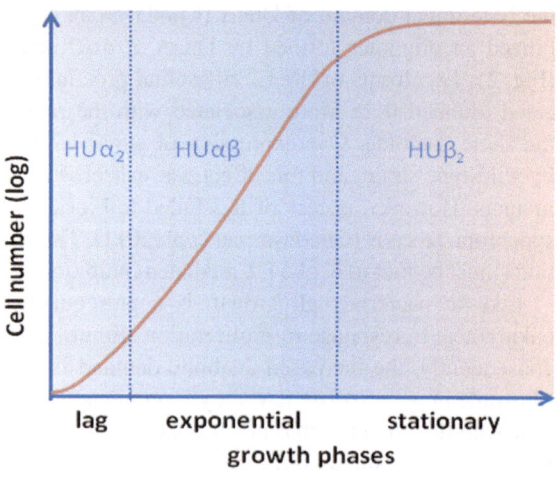

Fig. 3 HU composition changes according to growth phase. Only α subunit interacts with DnaA and enhances oligomer stability

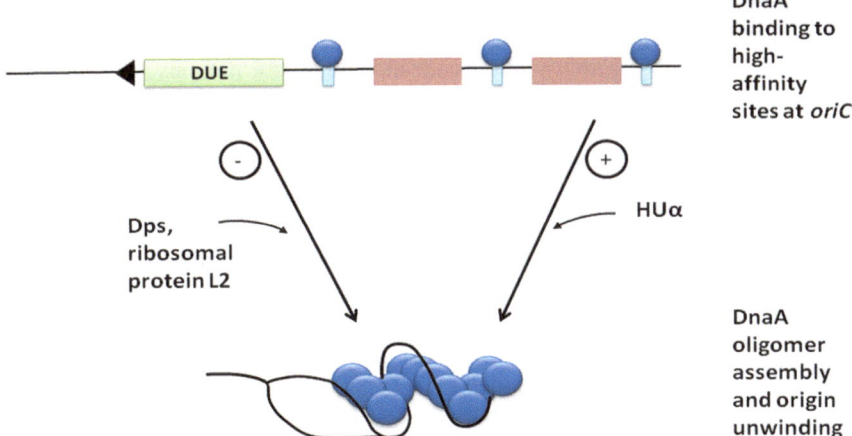

Fig. 4 Interaction of DnaA with nucleoid-associated proteins (HUα, Dps) and the ribosomal L2 protein regulates the formation of DnaA oligomer and subsequent origin unwinding. DnaA (*blue spheres*) binds to high-affinity sites (*light blue rectangles*) for most of the cell cycle. During the initiation of DnaA replication, it occupies low-affinity sites (shown as *pink panels*), which results in the formation of a helical DnaA filament and DNA unwinding at the DUE element. Position and directionality of transcription from the *gidA* promoter were depicted by a *black arrow*

chromosomal DNA non-specifically and stimulates strand opening at *oriC* in vitro. It was shown that the α dimer of HU has a stronger effect than other forms of the protein. The α subunit of HU interacts with the N-terminal region of DnaA and stabilizes the DnaA oligomer bound to *oriC* (Fig. 4). These observations led to the proposal that DnaA interacts with the α dimer or the αβ heterodimer, depending on their cellular abundance, to recruit the respective form of HU to *oriC*. It was suggested that this interaction regulates *oriC* activity in a growth phase-dependent manner (Chodavarapu et al. 2008a). In addition, Dps, a nucleoid-associated protein, which is expressed in response to oxidative stress and protects the chromosome from hydroxyl radical-mediated damage (for a review, see Zeth et al. 2012), was shown to interact with the N-terminal region of DnaA. DnaA–Dps interaction impedes initiation of replication during stress conditions by blocking *oriC* unwinding (Chodavarapu et al. 2008b).

Interaction of DnaA with Acidic Components of Fluid Cellular Membrane Regulates Its Nucleotide-Bound Status and Binding to *OriC*

Biological membranes take an active part in a number of vital processes such as energy transformation, transport, signal transduction, motility, protein trafficking, and many others. It has become evident that membrane lipids do not only form a static scaffold for membrane proteins, but also affect their folding and structure and influence formation of functional complexes. Membrane lipid components also directly regulate activity and localization of many proteins (for a review, see Phillips et al. 2009; Marsh 2008; Bogdanov et al. 2014). In addition, a large body of evidence suggests that lipid composition of the membrane is heterogeneous (Vanounou et al. 2002, 2004; Mieleykovskaya and Dowhan 2009; Barak and Muchova 2013). Lipids form clusters and microdomains within the envelope, whereas their number and localization fluctuate with growth conditions and the cell cycle (Mozharov et al. 1985; Hiraoka et al. 1993), suggesting that the inhomogeneous and highly dynamic structure of the cellular membrane may play a global regulatory role in bacteria (Fishov and Norris 2012). In this light, it is of particular interest that acidic phospholipids affect the activity of the DnaA initiator protein by promoting exchange of bound nucleotides (ATP or ADP) and inhibiting the formation of the prereplication complex (for a review, see Saxena et al. 2013).

E. coli membrane, in majority, consists of various phospholipids, and among them, zwitterionic phosphatidylethanolamine (PE) constitutes 70 % of the envelope lipids, whereas acidic phosphatidylglycerol (PG) and cardiolipin (CL) make up for 20 and 10 %, respectively (Barak and Muchova 2013). Acidic phospholipids are synthesized through common metabolic pathway involving phosphatidylglycerol phosphate synthase A (*pgsA* gene product). Depletion of the pool of acidic phospholipids by mutation or downregulation of *pgsA* affects bacterial growth (Heacock and Dowhan 1987; Fingland et al. 2012). Cells bearing the *pgsA* gene under control of an inducible promoter grow in the presence of the inducer in the medium, but after its removal, the growth is continued for several generations until a threshold amount of acidic phospholipids is reached, when cells undergo a growth arrest. If expression of phosphatidylglycerol phosphate synthase A is induced again in such cells, DNA replication is resumed prior to restoration of growth, which suggests that a specific cell cycle arrest might have occurred in the absence of sufficient amount of acidic phospholipids (Fingland et al. 2012).

Deleterious effects of the decreased content of acidic phospholipids may be overcome by changes in the DNA replication control. Namely, growth arrest of *pgsA* mutants is bypassed by a secondary mutation in gene encoding RNase H (*rnhA*) (Xia and Dowhan 1995). Lack of RNase H activity results in the formation of persistent RNA–DNA hybrids which serve as alternative sites for the initiation of chromosomal DNA replication, in a process called constitutive stable DNA replication (cSDR) that is dependent on RecA but independent of DnaA activity (von Meyenburg et al. 1987). In addition, overexpression of certain DnaA protein variants, bearing changes in the regions identified as responsible for the interaction with the membrane or DNA, suppresses the *pgsA* mutant growth defect (Zheng et al. 2001). In particular, production of the mutant DnaA L366-K protein, which on its own is unable to form fully functional prereplication complexes (Saxena et al. 2011), allows for the growth of acidic phospholipid-depleted cells in the presence of wild-type DnaA (Li et al. 2005; Zheng et al. 2001).

As mentioned in the preceding chapters, initiation of DNA replication at *oriC* requires attaining a certain threshold of ATP–DnaA molecules. During the elongation step, RIDA process and IHF-assisted interaction with the *datA* region lower the initiation potential of DnaA by stimulating hydrolysis of the bound ATP (Kasho and Katayama 2013). Thus, prior to the initiation of the next round of DNA replication, proportion of ATP–DnaA to ADP–DnaA rises (Kurokawa et al. 1999), which can occur through *de novo* DnaA synthesis or rejuvenation of ATP–DnaA pool by exchange of the nucleotide. The latter process is stimulated by the acidic phospholipids, which promote the release of the tightly bound ADP. In vitro, *oriC*-bound DnaA–ADP exchanges ADP for ATP in the presence of CL or PG and excess of ATP (Yung and Kornberg 1988; Crooke et al. 1992; Catsuma et al. 1993). CL is the most efficient in promoting DnaA–ADP rejuvenation, whereas zwitterionic phospholipids, such as PE, fail to stimulate nucleotide exchange (Crooke et al. 1992; Catsuma et al. 1993). In addition to their influence on nucleotide binding by the replication initiator, acidic phospholipids inhibit the interaction of DnaA with *oriC* (Crooke at al. 1992). Interestingly, when DnaA is exposed first to acidic phospholipids, its binding to the *origin* is abolished, whereas *oriC*-bound DnaA complexes are stable in the presence of these membrane components. This indicates that the process of ordered assembly of the pre-replication complex might be affected by the association of DnaA with membrane phospholipids (Crooke et al. 1992; Saxena et al. 2013). In addition to the presence of the acidic head group, phospholipid-mediated stimulation of ADP release by DnaA requires certain degree of membrane fluidity (Catsuma et al. 1993). Fatty acid components of the phospholipids have a strong impact on the DnaA–ADP rejuvenation, and only these containing unsaturated fatty acids are active in the stimulation of this process (Fralick and Lark 1973; Yung and Kornberg 1988; Catsuma et al. 1993). Membrane fluidity of *E. coli* varies with growth temperature (Marr and Ingraham 1963), which suggests that the impact of acidic phospholipids on the regulation of DnaA activity may be temperature dependent.

In situ counting of immunogold-labeled DnaA molecules allowed for the estimation that around 70 % of the cellular DnaA resides in the vicinity of the cell envelope (Newman and Crooke 2000). Recent studies employing the cell fractionation and detection of both native and fluorescent mCherry–DnaA fusion revealed that approximately 10 % of the replication initiator is stably associated with the inner membrane (Regev et al. 2012). This fraction was proposed to represent free DnaA available for the replication initiation, in contrast to the vast majority of molecules bound by chromosomal sequences, mainly *datA* (Regev et al. 2012). A region of DnaA forming amphipathic helices was identified as necessary for membrane binding (amino acids 354–372) (Garner et al. 1998; Garner and Crooke 1996; Yamaguchi et al. 1999). However, Regev and colleagues, based on the analysis of association of various DnaA fragments with the membrane fraction and modeling of electrostatic properties of the DnaA protein surface, proposed that hydrophobic patch along domains IIIa and IIIb of DnaA, opposite the ATP-binding pocket, is responsible for the interaction with membrane lipids, whereas amphipathic helices are essential for the intrinsic structure of the fragment, rather than membrane binding *per se* (Regev et al. 2012). The same authors suggested that such mode of association of the DnaA protein with membrane lipids indicates a mechanism of nucleotide exchange, which is common to other proteins regulated by relocation from the cytoplasm to the cell envelope, based on the conformational change of the protein, "straightening" of its structure and "lifting the lid" of the ATP-binding pocket (Regev et al. 2012). In domain IIIb, hydrophobic patch surrounds positively charged bulge composed mainly of Lys372 residue, which was previously shown to be indispensable for the CL-mediated release of ADP by DnaA (Hase et al. 1998; Makise et al. 2000; Regev et al. 2012). This confirms the earlier postulated importance of electrostatic interactions formed between DnaA and acidic phospholipids during the reactivation of the replication initiator protein (Kitchen et al. 1999). In addition, it was demonstrated that macromolecular crowding of DnaA on the membrane surface results in a rapid increase in the kinetics of nucleotide exchange (Aranovich et al. 2006). Thus, it was proposed that timing of the initiation of DNA replication is ensured by the rejuvenation of the initiator protein, mediated by continuously growing membrane domain, which upon reaching specific DnaA-to-phospholipid ratio catalyzes nucleotide exchange and hence restores initiation potential of DnaA. This way, initiation of DNA replication would be coupled to cellular mass (Aranovich et al. 2006; Regev et al. 2012).

Influence of DNA Replication on Cell Structure and Function

In the preceding chapters, we presented data on various cellular processes contributing to the regulation of DNA replication. In the following, we would like to focus on how activities of replication initiator proteins integrate various aspects of the cell cycle and how DNA replication process influences chromosomal and cellular structure.

Dysfunction of replication initiators has long been known to lead to pleiotropic phenotypes, both in eukaryotic and in prokaryotic cells. It was, however, not clear whether these effects were an indirect consequence of perturbed replication or an evidence for additional roles of these proteins, beyond the one played during the replication initiation. A large body of evidence has accumulated now, suggesting that replication initiator proteins take part in other processes, coordinating DNA replication with cell cycle and metabolism (for a review, see Scholefield et al. 2009). In *E. coli*, a role for DnaA as a transcription factor is well documented (Messer and Weigel 2003). DnaA can both repress and activate transcription. Although rules that govern the effect exerted by DnaA with respect to particular promoters are not known, it seems that repression of transcription involves formation of higher-order oligomeric structures (Olliver et al. 2010). As mentioned before, DnaA regulates transcription of its cognate gene, which was proposed to contribute to the coordination of DNA replication with cell cycle and growth rate (Polaczek and Wright 1990; Speck et al. 1999). In addition, it has been shown recently that temperature-dependent formation of specific complexes by DnaA at its own promoter region results in alterations of *dnaA* gene transcription level and takes part in adaptation of cell growth to different temperatures (Saggioro et al. 2013). Importantly, DnaA was also shown to control the levels of ribonucleotide reductase, by both activating and repressing transcription from *nrdAB* promoter as a function of its concentration (Olliver et al. 2010; Augustin et al. 1994, Gon et al. 2006). Nucleotide reductase, encoded by *nrdAB* operon, is an essential enzyme involved in the last step of synthesis of deoxynucleotides necessary for DNA replication. Imbalances in dNTP levels increase mutation rate and threaten genomic integrity (Wheeler et al. 2005; Mathews et al. 2006). It has been shown recently that the amount of ribonucleotide reductase produced is proportional to DnaA–ATP level; however, DnaA-dependent changes of the *nrdAB* promoter activity do not regulate the timing of *nrdAB* operon expression as a function of cell cycle (Olliver et al. 2010). Importantly, regulation of this operon by the DnaA protein was also observed in *B. subtilis* and *C. crescentus* (Goranov et al. 2005; Breier and Grossman 2009; Hottes et al. 2005). In these two organisms, DnaA—by regulating the level of transcription—coordinates DNA replication with cell division. In *B. subtilis*, DnaA negatively affects the expression of the *ftsL* gene, encoding membrane-bound protein involved in the recruitment of the cell division machinery (Bramkamp et al. 2006; Goranov et al. 2005; Breier and Grossman 2009). In *C. crescentus*, DnaA acts as a direct transcriptional activator of several promoters: *mipZ*—controlling expression of a spatial regulator of cell division (MipZ) (Hottes et al. 2005; Fernandez-Fernandez et al. 2011), *ftsZ*—regulating the production of the cell division protein (FtsZ) (Hottes et al. 2005; Kelly et al. 1998; Quardokus et al. 1996), and *gcrA*—driving transcription of the gene encoding master cell cycle regulator (GcrA) (Collier et al. 2006; Fernandez-Fernandez et al. 2011). In addition to several promoters of *E. coli* for whom transcriptional regulation by DnaA was investigated in more detail (*mioC*, *rpoH*, *uvrB*, *polA*, *glpD*, *fliC*), a study involving transcriptomic analysis of *dnaA46* temperature-sensitive mutant revealed that expression of 227 genes was changed more than 2-fold upon temperature upshift (Lobner-Olesen et al. 2008). Affected genes belonged to several functional categories, and among them are nucleotide synthesis, CCM, and fatty acid and phospholipid metabolism. Although changes of expression of many of these genes are most likely indirectly affected by the inactivation of DnaA, since examination of promoter regions of some of them did not reveal the presence of DnaA boxes (Lobner-Olesen et al. 2008), it cannot be excluded that DnaA regulon in *E. coli* encompasses more genes than have been characterized so far, taken also into account various modes of DnaA binding (Glinkowska et al. 2003; Ozaki et al. 2001).

DnaA has been implicated in the coordination of replication and metabolism also in *B. subtilis*. Studies on genome-wide DnaA interaction by ChIP–chip experiments and transcription profiling suggested that DnaA binds and regulates transcription from promoters of several genes involved in nucleotide and carbohydrate metabolism, iron homeostasis, and ribosome biogenesis (Ishikawa et al. 2007; Goranov et al. 2005; Breier and Grossman 2009). In this organism, DnaA is also responsible for the correlation of DNA replication with sporulation. It has been shown that there is a specific time window within *B. subtilis* cell cycle when sporulation can be triggered, presumably to ensure synthesis of two complete chromosomal copies before the onset of spore formation. Expression of the *sda* gene, responsible for the inhibition of sporulation, occurs in pulsatile manner during the cell cycle. Up-regulation of *sda* transcription coincides with the initiation of DNA replication and is coordinated by the active form of DnaA replication initiator. Together with intrinsic instability of the Sda protein, cell cycle-coupled expression prevents the formation of spores containing replicating chromosomes (Veening et al. 2009).

Involvement of DNA replication initiator proteins in coordinating chromosomal DNA synthesis with the cell cycle has also been recently shown in eukaryotic cells. To facilitate the accurate chromosome segregation, replicated chromosomes are held together by sister chromatid cohesion (SCC), until the metaphase-to-anaphase transition, when chromosomes become separated (for a review, see Nasmyth and Hering 2009). Two independent mechanisms of chromatid cohesion have been identified, and proteins of origin recognition complex (ORC) have been implicated in both of them (for a review, see Sasaki and Gilbert 2007). The first is based on cohesin—a protein that encircles the paired DNA molecules. In *Xenopus* egg extracts, prereplication complex formation is required for cohesin loading, and in *Drosophila,* cohesion preferentially colocalizes with ORC (Takahashi et al. 2004; McAlpine et al. 2010; Sherwood at al. 2010). A newly identified mechanism of SCC has been described in budding yeast. Cells depleted for ORC2 during the G1 phase continued through S phase (ORC is dispensable for pre-RC maintenance) but were arrested in mitosis due to spindle checkpoint activation. Arrested cells displayed precocious chromatid separation, although cohesin complexes were recruited normally (Shimada & Gasser 2007). These results corroborated previous study, which showed that some *orc* gene mutations are synthetically lethal with cohesion complex mutants (Suter et al. 2004).

The role of ORC in cohesin complexes' localization may be linked to its function in chromatin silencing, as heterochromatin is necessary for SCC in many systems (Chang et al. 2005; Chen et al. 2012). It was shown in budding yeast that separable domains of ORC proteins were involved in the DNA replication and establishment of silent chromatin at mating-type loci (Bell 2002; Ozaydin et al. 2010). ORC1 was shown to recruit chromatin-silencing protein Sir1 via direct protein–protein interactions (Hou et al. 2005; Hsu et al. 2005). This seems to be an evolutionary conserved role of ORC proteins, as they were shown to interact directly with HP1 protein involved in heterochromatin organization in *Drosophila* and mammals (Pak et al. 1997; Auth et al. 2006; Prasanth et al. 2010, Chakraborty et al. 2011). In addition to influencing chromatin structure, ORC may also mediate localized gene repression, as suggested by the results of studies showing that a class of genes becomes induced by *orc2-1* mutation in *S. cerevisiae* (Ramachandran et al. 2006). It has been suggested that coevolution of replication and chromatin-silencing functions of ORC proteins may have been driven by the need of different segments of large eukaryotic chromosomes to be replicated at different times during S phase, and ORC role in silencing may contribute to the distribution of early and late replicating sites, as *origins* with the strongest ORC binding activity function poorly as replication initiation sites and strongly as silencers (for a review, see Sasaki and Gilbert 2007). Recently, a study employing *orc2-1* mutant, in which only sites that bind ORC tightly remain fully occupied, showed that ORC-interacting sites were comprised of protein-coding regions of highly transcribed metabolic genes (ORF-ORC). In contrast to the ORC-silencer paradigm, transcriptional activation promoted ORC association with these genes. Remarkably, ORF–ORC genes were enriched in proximity to *origin*s of replication and, in several instances, were transcriptionally regulated by these origins. Taken together, these results suggested a surprising connection between ORC, replication origins, and cellular metabolism (Shor et al. 2009).

ORC components have also been implicated in chromosome segregation and condensation. It was suggested that ORC proteins might take part in coupling of completion of chromosome synthesis with chromosome condensation and spindle attachment to kinetochores (Sasaki and Gilbert 2007). ORC1 and ORC4 colocalize with kinetochore in the absence of replicative helicase in fission yeast (Hayashi et al. 2007). In human cells, ORC2 is released from chromatin except kinetochores at the S phase (Hayashi et al. 2007), and ORC6 was also localized to kinetochores during mitosis (Prasanth et al. 2002).

In *Drosophila*, defects in ORC2, ORC4, or ORC5 lead to abnormally condensed chromosomes, suggesting a role of these proteins in the condensation process (Pflumm and Botcham 2001). However, similarly to the role of ORC in the function of centromeres, a direct link between them and the respective processes is missing and aberrations in chromosome condensation and segregation observed in ORC mutants may stem from perturbed replication or defects in heterochromatin assembly.

Another role in addition to participation in DNA replication has been ascribed to ORC6, which was postulated to coordinate DNA synthesis and cytokinesis. ORC6 localizes to the cell membrane (Chesnokov et al. 2001) in *Drosophila* and has been shown to interact with proteins involved in cytokinesis both in *Drosophila* and in mammals (Huijbregts et al. 2009; Prasanth et al. 2002). Reduction of ORC6 synthesis leads to the appearance of cells that completed mitosis without cytokinesis.

Taken together, results obtained with prokaryotic and eukaryotic cells suggest that replication initiation proteins universally play roles additional to the ones in pre-RC assembly. Their additional functions help to coordinate various cellular activities with DNA replication.

In addition to the participation of replication proteins in the control of various cellular processes, DNA dynamics during the DNA replication may influence DNA transactions and protein distribution in the cell. Liu and Wang have shown that transcribing RNA polymerase generates waves of supercoiling, positive in front of the protein complex and negative in its wake (Liu and Wang 1987). DNA supercoiling has been implicated as an important regulatory factor in transcription, recombination, and DNA replication (for a review, see Fogg et al. 2012; Baranello et al. 2012; Travers and Muskhelishvili 2005). Until recently, the prevailing view was that the twin-domain model is not fully applicable to the advancing replisome, which—due to discontinuous nature of the replication process—produces mainly positive supercoiling in front of the complex (for a review, see Yu and Dröge 2014). The degree of DNA topological changes behind the replisome was uncertain; however, recent reports provided evidence that leading strand becomes negatively supercoiled in the wake of replication machinery. Studies of the replication complexes assembled in vitro using rolling-circle replication substrate have proven that negative helical tension, created by *E. coli* replisome on the leading strand, causes dissociation of the lagging strand polymerase (Kurth et al. 2013). Results of experiments conducted in yeasts on the removal of ribonucleotides from nascent DNA identified topoisomerase I as important component of the nucleotide removal pathway. Eukaryotic topoisomerase I preferentially recognizes superhelical DNA, and thus, the results provide a hint that negative superhelical tension builds up in the leading duplex, resulting in the structural changes of DNA–RNA hybrid, cleaved by Topo I (Williams et al. 2013). Promoters of many genes, both in eukaryotic and in prokaryotic cells, are controlled by DNA supercoiling (Travers and Muskhelishvili 2005; Blot et al. 2006; Lavelle 2014). Therefore, negative superhelicity generated by the replisome may constitute an important factor influencing gene expression and—on the evolutionary scale—also genome organization. In fact, this hypothesis was previously implied in two studies. The spatial organization of transcription programs during different growth phases was proposed to be coupled to DNA replication process, and the supercoiling gradient generated both by replication and by the asymmetric distribution of gyrase target sites along the Ori–Ter axis (Sobetzko et al. 2012, 2013). The findings presented in these works suggest that in *E. coli* chromosome, not only the spatial gene order, but also the physical properties of their coding sequences correspond with spatiotemporal pattern of superhelicity, associated with DNA replication. In particular, mapping of the physical properties of *E. coli* chromosomal DNA revealed that a gradient of DNA melting energy, closely corresponding to the gradient of gyrase binding sites, extends from *origin* to *terminus*, whereby the DNA sequences in the Ori end of the chromosome region have higher average melting energy than those around Ter. Growth cycle-resolved transcriptomics, subsequent mapping of temporal patterns of gene expression in the *E. coli* chromosome, and analysis of spatial correlation between functional and regulatory features of expressed genes demonstrated that utilization of higher melting energy sequences increases gradually during exponential phase in parallel with high oxygen consumption. These genes characterized by higher melting energy sequences are located in spatial proximity to replication *origin*, have preferential positioning on the leading strand and correspond to previously identified genes activated by high negative DNA superhelicity. Opposite correlation was observed for genes closer to the terminus region, found to utilize lower melting energy sequences. These genes were preferentially activated during entry into the stationary phase, on decreased oxygen consumption and required DNA relaxation.

The spatiotemporal changes in the gene expression most likely reflect structural reorganization of chromosome during growth, whereas the correlation of physical features of transcribed sequences with their positioning along the trajectory of replisomes suggests that DNA replication and the gene expression program might be integrated by chromosome organization and reciprocally, that nucleoid structure is determined by the interplay of these processes in response to availability of nutrients and oxygen. Results of these works strongly suggest that sequence organization of the genome and changes in chromosomal DNA supercoiling coordinate the global gene expression with DNA replication (Sobetzko et al. 2012, 2013).

A question also arises whether changes in the replication organization in response to environmental conditions can be found which could determine chromosome structure, and in turn, could it play a role in orchestrating global transcription pattern with changing metabolic state of the cell. One obvious difference in the replication pattern, which can be dissected under feast versus famine conditions, is that the fast growing *E. coli* cells can support multifork replication. SeqA protein, which binds hemimethylated DNA and is responsible for *origin* sequestration (for a review, see Waldminghaus and Skarstad 2009), was also shown to be involved in the organization of replication forks into conspicuous structures, which can be microscopically visualized as SeqA foci (Molina and Skarstad 2004). The number of foci and the extent of their colocalization with the replication fork depend on the growth rate, and as no increase in their number is observed during replication initiation, new replication forks were proposed to be recruited to existing structures (Morigen et al. 2009; Fossum et al. 2007). Recently, SeqA binding was demonstrated to follow the emergence of hemimethylated DNA behind the moving replication forks (Waldminghaus et al. 2012). However, the extent of this binding was dependent on the growth rate. During the fast growth, SeqA disaggregated from the old replication forks as they were approaching the terminus region. This resulted in the low efficiency of binding of SeqA in the proximity of terminus in fast, relative to slow, growing cells (Waldminghaus et al. 2012).

SeqA affects global chromosome structure, and cells devoid of this protein have increased level of DNA superhelicity (Skarstad et al. 2001), display changes in chromosome morphology, and have segregation problems (Kang et al. 2003). Transcription pattern is also changed in cells lacking SeqA, which is accompanied by severe growth impairment during high but not low growth rate (Lobner-Olesen et al. 2003). Furthermore, SeqA has been described also to act as a transcription factor (Slominska et al. 2001, 2003). All these observations make SeqA a likely candidate for a factor integrating chromosome structure with the demands of replication and transcription process depending on growth conditions. Strikingly also, mutants devoid of SeqA do not undergo growth arrest after the induction of ppGpp synthesis, but the nature of the interplay between SeqA and stringent response remains unknown (Ferullo and Lovett 2008). Recent study on factors determining global nucleoid structure in *E. coli*, employing chromosome conformation capture technology, revealed a high degree of clustering of SeqA binding sites, primarily locating in the Ori domain in exponentially growing cells, which confirms the important role of this protein in linking DNA replication with chromosome structure (Cagliero et al. 2013). It is also worth noting that results of recent studies on *C. crescentus* demonstrated that localization and number of polyphosphate granules, compounds with regulatory functions, and Ppk1 enzyme responsible for their synthesis are governed by replication and chromosome segregation processes (Henry and Crosson 2013). These findings underscore the role of chromosome dynamics associated with DNA replication in proteins' localization and organization of cellular processes.

DNA synthesis affects gene expression in bacteria also through replication-associated gene dosage. Bidirectional replication, especially in fast growing bacteria, results in transient overrepresentation of genes close to the *origin* of replication. This gene dosage effect was also suggested to be an important force shaping organization of bacterial genomes. Gene dosage effects have been shown to strongly constrain position in the vicinity of *origin* of genes involved in transcription and translation, but not other highly transcribed genes (Couturier and Rocha 2006; Rocha 2004).

As transcription and DNA replication take place on the same template, and in bacteria, transcription elongation rate is 12–30 times slower than that of DNA synthesis, frequent encounters of transcription and replication complexes must occur, which may lead to replication fork stalling or collapse and threaten genomic integrity. Therefore, multiple mechanisms exist to resolve replication–transcription conflicts. Their description is beyond the scope of this work; however, we recommend excellent reviews covering this topic in both bacterial and eukaryotic cells (Pomerantz and O'Donnell 2010; Lin and Pasero 2012; Bermejo et al. 2012). Nevertheless, we would like to emphasize that preventing conflicts between replication and transcription is another source of evolutionary pressure on genome layout (for a review, see Kepes et al. 2012). As codirectional encounters of transcription and replication machineries are less deleterious than head-on ones, highly expressed genes tend to be transcribed codirectionally with replication in numerous species (Rocha and Danachin 2003; Rocha 2008). Such organization is observed both in *B. subtilis* and in *E. coli,* for all rRNA operons (Guy and Roten 2004). Similarly, essential genes are enriched on the leading strand (Rocha et al. 2008). Since probability of collision between transcription and replication complexes increases with gene length, longer genes are also preferentially transcribed on the leading strand (Omont and Kepes 2004). Thus, replication-associated effects impose evolutionary pressure on genome organization and have to be taken into account during potential construction of synthetic organisms.

Conclusions

A wealth of data suggests that coordination of DNA replication with growth conditions may involve direct participation of proteins with primary function in other vital cellular processes, as well as metabolites, DNA structure, and cell envelope. Reciprocally, replication factors may take part in control of other processes, and replication-associated changes in DNA topology may be influential in the regulation of cellular functions. Regulation of cell cycle may involve the formation of macromolecular assemblies which integrate extra- and intracellular signals and allow their transformation to fewer outcomes that can be interpreted by the cell. Studies devoted to these and other aspects of cell cycle regulation, described in this work, will bring much better understanding of the basis of functioning of living systems.

Acknowledgments The authors acknowledge support from National Science Center (project grants: 2011/02/A/NZ1/00009 and 2012/04/M/NZ2/00122).

References

Abe Y, Jo T, Matsuda Y, Matsunaga C, Katayama T, Ueda T (2007) Structure and function of DnaA N-terminal domains: specific sites and mechanisms in inter-DnaA interaction and in DnaB helicase loading on oriC. J Biol Chem 282:17816–17827

Aranovich A, Gdalevsky GY, Cohen-Luria R, Fishov I, Parola AH (2006) Membrane-catalyzed nucleotide exchange on DnaA. Effect of surface molecular crowding. J Biol Chem 281:12526–12534

Atlung T, Løbner-Olesen A, Hansen FG (1987) Overproduction of DnaA protein stimulates initiation of chromosome and minichromosome replication in Escherichia coli. Mol Gen Genet 206:51–59

Atlung T (1984) Allele-specific suppression of dnaA(Ts) mutations by rpoB mutations in Escherichia coli. Mol Gen Genet 197:125–128

Auth T, Kunkel E, Grummt F (2006) Interaction between HP1alpha and replication proteins in mammalian cells. Exp Cell Res 312:3349–3359

Ball CA, Osuna R, Ferguson KC, Johnson RC (1992) Dramatic changes in Fis levels upon nutrient upshift in Escherichia coli. J Bacteriol 174:8043–8056

Barák I, Muchová K (2013) The role of lipid domains in bacterial cell processes. Int J Mol Sci 14:4050–4065

Baranello L, Levens D, Gupta A, Kouzine F (2012) The importance of being supercoiled: how DNA mechanics regulate dynamic processes. Biochim Biophys Acta 1819:632–638

Baranska S, Gabig M, Wegrzyn A, Konopa G, Herman-Antosiewicz A, Hernandez P, Schvartzman JB, Helinski DR, Wegrzyn G (2001) Regulation of the switch from early to late bacteriophage lambda DNA replication. Microbiology 147:535–547

Barańska S, Glinkowska M, Herman-Antosiewicz A, Maciąg-Dorszyńska M, Nowicki D, Szalewska-Pałasz A, Węgrzyn A, Węgrzyn G (2013) Replicating DNA by cell factories: roles of central carbon metabolism and transcription in the control of DNA replication in microbes, and implications for understanding this process in human cells. Microb Cell Factor 12:55

Bates DB, Boye E, Asai T, Kogoma T (1997) The absence of effect of gid or mioC transcription on the initiation of chromosomal replication in Escherichia coli. Proc Natl Acad Sci USA 94:12497–12502

Becker NA, Kahn JD, Maher LJ 3rd (2005) Bacterial repression loops require enhanced DNA flexibility. J Mol Biol 349:716–730

Bell SP (2002) The origin recognition complex: from simple origins to complex functions. Genes Dev 16:659–672

Bermejo R, Lai MS, Foiani M (2012) Preventing replication stress to maintain genome stability: resolving conflicts between replication and transcription. Mol Cell 45:710–718

© The Author(s) 2015

M. Glinkowska et al., *DNA Replication Control in Microbial Cell Factories*, SpringerBriefs in Microbiology, DOI 10.1007/978-3-319-10533-8

37

Blaesing F, Weigel C, Welzeck M, Messer W (2000) Analysis of the DNA-binding domain of *Escherichia coli* DnaA protein. Mol Microbiol 36:557–569

Blot N, Mavathur R, Geertz M, Travers A, Muskhelishvili G (2006) Homeostatic regulation of supercoiling sensitivity coordinates transcription of the bacterial genome. EMBO Rep 7:710–715

Bogdanov M, Dowhan W, Vitrac H (2014) Lipids and topological rules governing membrane protein assembly. Biochim Biophys Acta. (in press), doi:10.1016/j.bbamcr.2013.12.007

Booker BM, Deng S, Higgins NP (2010) DNA topology of highly transcribed operons in Salmonella enterica serovar Typhimurium. Mol Microbiol 78:1348–1364

Boye E, Nordström K (2003) Coupling the cell cycle to cell growth. EMBO Rep 4:757–760

Bramkamp M, Weston L, Daniel RA, Errington J (2006) Regulated intramembrane proteolysis of FtsL protein and the control of cell division in Bacillussubtilis. Mol Microbiol 62:580–591

Breier AM, Grossman AD (2009) Dynamic association of the replication initiator and transcription factor DnaA with the Bacillus subtilis chromosome during replication stress. J Bacteriol 191:486–493

Breier AM, Weier HU, Cozzarelli NR (2005) Independence of replisomes in *Escherichia coli* chromosomal replication. Proc Natl Acad Sci U S A 102:3942–3947

Bremer H, Churchward G (1985) Initiation of chromosome replication in *Escherichia coli* after induction of dnaA gene expression from a lac promoter. J Bacteriol 164:922–924

Bremer H, Dennis PP (1996) Modulation of chemical composition and other parameters of the cell by growth rate. In: Neidhardt FC (ed) *Escherichia coli* and Salmonella, vol 2. ASM Press, Washington, pp.1553–1569

Browning DF, Grainger DC, Busby SJ (2010) Effects of nucleoid-associated proteins on bacterial chromosome structure and gene expression. Curr Opin Microbiol 13:773–780

Butland G, Peregrín-Alvarez JM, Li J, Yang W, Yang X, Canadien V, Starostine A, Richards D, Beattie B, Krogan N, Davey M, Parkinson J, Greenblatt J, Emili A (2005) Interaction network containing conserved and essential protein complexes in *Escherichia coli*. Nature 433:531–537

Cabrera JE, Cagliero C, Quan S, Squires CL, Jin DJ (2009) Active transcription of rRNA operons condenses the nucleoid in *Escherichia coli*: examining the effect of transcription on nucleoid structure in the absence of transertion. J Bacteriol 191:4180–4185

Cagliero C, Grand RS, Jones MB, Jin DJ, O'Sullivan JM (2013) Genome conformation capture reveals that the *Escherichia coli* chromosome is organized by replication and transcription. Nucleic Acids Res 41:6058–6071

Campos M, Jacobs-Wagner C (2013) Cellular organization of the transfer of genetic information. Curr Opin Microbiol 16:171–176

Cassler MR, Grimwade JE, Leonard AC (1995) Cell cycle-specific changes in nucleoprotein complexes at a chromosomal replication origin. EMBO J 14:5833–5841

Castuma CE, Crooke E, Kornberg A (1993) Fluid membranes with acidic domains activate DnaA, the initiator protein of replication in *Escherichia coli*. J Biol Chem 268:24665–24668

Celler K, Koning RI, Koster AJ, van Wezel GP (2013) Multidimensional view of the bacterial cytoskeleton. J Bacteriol. 195:1627–1636

Chakraborty A, Shen Z, Prasanth SG (2011) "ORCanization" on heterochromatin: linking DNA replication initiation to chromatin organization. Epigenetics 6:665–670

Chang CR, Wu CS, Hom Y, Gartenberg MR (2005) Targeting of cohesin by transcriptionally silent chromatin. Genes Dev 19:3031–3042

Chen Z, McCrosky S, Guo W, Li H, Gerton JL (2012) A genetic screen to discover pathways affecting cohesin function in Schizosaccharomyces pombe identifies chromatin effectors. G3 (Bethesda) 2: 1161–1168

Chesnokov I, Remus D, Botchan M (2001) Functional analysis of mutant and wild-type Drosophila origin recognition complex. Proc Natl Acad Sci U S A 98:11997–12002

Chiaramello AE, Zyskind JW (1990) Coupling of DNA replication to growth rate in *Escherichia coli*: a possible role for guanosine tetraphosphate. J Bacteriol 172:2013–2019

Chodavarapu S, Felczak MM, Kaguni JM (2011) Two forms of ribosomal protein L2 of *Escherichia coli* that inhibit DnaA in DNA replication. Nucleic Acids Res 39:4180–4191

Chodavarapu S, Gomez R, Vicente M, Kaguni JM (2008a) *Escherichia coli* Dps interacts with DnaA protein to impede initiation: a model of adaptive mutation. Mol Microbiol 67:1331–1346

Chodavarapu S, Felczak MM, Yaniv JR, Kaguni JM (2008b) *Escherichia coli* DnaA interacts with HU in initiation at the E. coli replication origin. Mol Microbiol 67:781–792

Claret L, Rouviere-Yaniv J (1997) Variation in HU composition during growth of *Escherichia coli*: the heterodimer is required for long term survival. J Mol Biol 273:93–104

Collier J, Murray SR, Shapiro L (2006) DnaA couples DNA replication and the expression of two cell cycle master regulators. EMBO J 25:346–356

Cooper S, Helmstetter CE (1968) Chromosome replication and the division cycle of *Escherichia coli* B/r. J Mol Biol 31:519–540

Couturier E, Rocha EP (2006) Replication-associated gene dosage effects shape the genomes of fast-growing bacteria but only for transcription and translation genes. Mol Microbiol 59:1506–1518

Crooke E, Castuma CE, Kornberg A (1992) The chromosome origin of *Escherichia coli* stabilizes DnaA protein during rejuvenation by phospholipids. J Biol Chem 267:16779–16782

Czapla L, Grosner MA, Swigon D, Olson WK (2013) Interplay of protein and DNA structure revealed in simulations of the lac operon. PLoS One 8:e56548

Dai RP, Yu FX, Goh SR, Chng HW, Tan YL, Fu JL, Zheng L, Luo Y (2008) Histone 2B (H2B) expression is confined to a proper NAD+/NADH redox status. J Biol Chem 283:26894–26901

Deisenroth C, Zhang Y (2010) Ribosome biogenesis surveillance: probing the ribosomal protein-Mdm2-p53 pathway. Oncogene 29:4253–4260

Deisenroth C, Zhang Y (2011) The ribosomal protein-mdm2-p53 pathway and energy metabolism: bridging the gap between feast and famine. Genes Cancer 2:392–403

DeNapoli J, Tehranchi AK, Wang JD (2013) Dose-dependent reduction of replication elongation rate by (p)ppGpp in *Escherichia coli* and *Bacillus subtilis*. Mol Microbiol 88:93–104

Dennis PP, Bremer H (1974) Macromolecular composition during steady-state growth of *Escherichia coli* B-r. J Bacteriol 119:270–281

Dervyn E, Suski C, Daniel R, Bruand C, Chapuis J, Errington J, Jannière L, Ehrlich SD (2001) Two essential DNA polymerases at the bacterial replication fork. Science 294:1716–1719

Donachie WD (1968) Relationship between cell size and time of initiation of DNA replication. Nature 219:1077–1079

Donati G, Montanaro L, Derenzini M (2012) Ribosome biogenesis and control of cell proliferation: p53 is not alone. Cancer Res 72:1602–1607

Dorman CJ (2013) Genome architecture and global gene regulation in bacteria: making progress towards a unified model? Nat Rev Microbiol 11:349–355

Du YC, Stillman B (2002) Yph1p, an ORC-interacting protein: potential links between cell proliferation control, DNA replication, and ribosome biogenesis. Cell 109:835–848

Duderstadt KE, Berger JM (2008) AAA+ ATPases in the initiation of DNA replication. Crit Rev Biochem Mol Biol 43:163–187

Duderstadt KE, Chuang K, Berger JM (2011) DNA stretching by bacterial initiators promotes replication origin opening. Nature 478:209–213

Duderstadt KE, Mott ML, Crisona NJ, Chuang K, Yang H, Berger JM (2010) Origin remodeling and opening in bacteria rely on distinct assembly states of the DnaA initiator. J Biol Chem 285:28229–28239

Erzberger JP, Mott ML, Berger JM (2006) Structural basis for ATP-dependent DnaA assembly and replication-origin remodeling. Nat Struct Mol Biol 13:676–683

Erzberger JP, Pirruccello MM, Berger JM (2002) The structure of bacterial DnaA: implications for general mechanisms underlying DNA replication initiation. EMBO J 21:4763–4773

Felczak MM, Kaguni JM (2004) The box VII motif of *Escherichia coli* DnaA protein is required for DnaA oligomerization at the E. coli replication origin. J Biol Chem 279:51156–51162

Fernandez-Fernandez C, Gonzalez D, Collier J (2011) Regulation of the activity of the dual-function DnaA protein in Caulobacter crescentus. PLoS One 6:e26028

Ferreira E, Giménez R, Aguilera L, Guzmán K, Aguilar J, Badia J, Baldomà L (2013) Protein interaction studies point to new functions for *Escherichia coli* glyceraldehyde-3-phosphate dehydrogenase. Res Microbiol 164:145–154

Ferullo DJ, Lovett ST (2008) The stringent response and cell cycle arrest in *Escherichia coli*. PLoS Genet 4:e1000300

Fingland N, Flåtten I, Downey CD, Fossum-Raunehaug S, Skarstad K, Crooke E (2012) Depletion of acidic phospholipids influences chromosomal replication in *Escherichia coli*. Microbiologyopen 1:450–466

Fishov I, Norris V (2012) Membrane heterogeneity created by transertion is a global regulator in bacteria. Curr Opin Microbiol 15:724–730

Flåtten I (2009) Morigen, Skarstad K. DnaA protein interacts with RNA polymerase and partially protects it from the effect of rifampicin. Mol Microbiol 71:1018–1030

Foffi G, Pastore A, Piazza F, Temussi PA (2013) Macromolecular crowding: chemistry and physics meet biology (Ascona, Switzerland, 10–14 June 2012). Phys Biol 10:040301

Fogg JM, Randall GL, Pettitt BM, de Sumners WL, Harris SA, Zechiedrich L (2012) Bullied no more: when and how DNA shoves proteins around. Q Rev Biophys 45:257–299

Fossum S, Crooke E, Skarstad K (2007) Organization of sister origins and replisomes during multifork DNA replication in *Escherichia coli*. EMBO J 26:4514–4522

Fralick JA, Lark KG (1973) Evidence for the involvement of unsaturated fatty acids in initiating chromosome replication in *Escherichia coli*. J Mol Biol 80:459–475

Fujikawa N, Kurumizaka H, Nureki O, Terada T, Shirouzu M, Katayama T, Yokoyama S (2003) Structural basis of replication origin recognition by the DnaA protein. Nucleic Acids Res 31:2077–2086

Garner J, Durrer P, Kitchen J, Brunner J, Crooke E (1998) Membrane-mediated release of nucleotide from an initiator of chromosomal replication, *Escherichia coli* DnaA, occurs with insertion of a distinct region of the protein into the lipid bilayer. J Biol Chem 273:5167–5173

Garner J, Crooke E (1996) Membrane regulation of the chromosomal replication activity of E. coli DnaA requires a discrete site on the protein. EMBO J 15:3477–3485

Gatenby RA, Gillies RJ (2004) Why do cancers have high aerobic glycolysis? Nat Rev Cancer 4:891–899

Gawel D, Jonczyk P, Fijalkowska IJ, Schaaper RM (2011) Mutator of *Escherichia coli*: effects of the τ subunit of the DNA polymerase III holoenzyme on chromosomal DNA replication fidelity. J Bacteriol 193:296–300

Gille H, Egan JB, Roth A, Messer W (1991) The FIS protein binds and bends the origin of chromosomal DNA replication, oriC, of *Escherichia coli*. Nucleic Acids Res 19:4167–4172

Glinkowska M, Majka J, Messer W, Wegrzyn G (2003) The mechanism of regulation of bacteriophage lambda pR promoter activity by *Escherichia coli* DnaA protein. J Biol Chem 278:22250–22256

Gon S, Camara JE, Klungsøyr HK, Crooke E, Skarstad K, Beckwith J (2006) A novel regulatory mechanism couples deoxyribonucleotide synthesis and DNA replication in *Escherichia coli*. EMBO J 25:1137–1147

Goranov AI, Katz L, Breier AM, Burge CB, Grossman AD (2005) A transcriptional response to replication status mediated by the conserved bacterial replication protein DnaA. Proc Natl Acad Sci USA 102:12932–12937

Govindarajan S, Nevo-Dinur K, Amster-Choder O (2012) Compartmentalization And Spatio-Temporal Organization Of Macromolecules In Bacteria. FEMS Microbiol Rev 36:1005–1022

Grant MA, Saggioro C, Ferrari U, Bassetti B, Sclavi B (2011) Cosentino Lagomarsino M. DnaA and the timing of chromosome replication in *Escherichia coli* as a function of growth rate. BMC Syst Biol 5:201

Guy L, Roten CA (2004) Genometric analyses of the organization of circular chromosomes: a universal pressure determines the direction of ribosomal RNA genes transcription relative to chromosome replication. Gene 340:45–52

Guzmán EC, Caballero JL, Jiménez-Sánchez A (2002) Ribonucleoside diphosphate reductase is a component of the replication hyperstructure in *Escherichia coli*. Mol Microbiol 43:487–495

Hansen FG, Christensen BB, Atlung T (2007) Sequence characteristics required for cooperative binding and efficient in vivo titration of the replication initiator protein DnaA in E. coli. J Mol Biol 367:942–952

Hansen FG, Atlung T, Braun RE, Wright A, Hughes P, Kohiyama M (1991) Initiator (DnaA) protein concentration as a function of growth rate in *Escherichia coli* and Salmonella typhimurium. J Bacteriol 173:5194–5199

Hase M, Yoshimi T, Ishikawa Y, Ohba A, Guo L, Mima S, Makise M, Yamaguchi Y, Tsuchiya T, Mizushima T (1998) Site-directed mutational analysis for the membrane binding of DnaA protein. Identification of amino acids involved in the functional interaction between DnaA protein and acidic phospholipids. J Biol Chem 273:28651–28656

Hasnain G, Waller JC, Alvarez S, Ravilious GE, Jez JM, Hanson AD (2012) Mutational analysis of YgfZ, a folate-dependent protein implicated in iron/sulphur cluster metabolism. FEMS Microbiol Lett 326:168–172

Hayashi M, Katou Y, Itoh T, Tazumi A, Yamada Y, Takahashi T, Nakagawa T, Shirahige K, Masukata H (2007) Genome-wide localization of pre-RC sites and identification of replication origins in fission yeast. EMBO J 26:1327–1339

Hayashi M, Ogura Y, Harry EJ, Ogasawara N, Moriya S (2005) Bacillus subtilis YabA is involved in determining the timing and synchrony of replication initiation. FEMS Microbiol Lett 247:73–79

He H, Lee MC, Zheng LL, Zheng L, Luo Y (2012) Integration of the metabolic/redox state, histone gene switching, DNA replication and S-phase progression by moonlighting metabolic enzymes. Biosci Rep 33:e00018

Heacock PN, Dowhan W (1987) Construction of a lethal mutation in the synthesis of the major acidic phospholipids of *Escherichia coli*. J Biol Chem 262:13044–13049

Heichlinger A, Ammelburg M, Kleinschnitz EM, Latus A, Maldener I, Flärdh K, Wohlleben W, Muth G (2011) The MreB-like protein Mbl of *Streptomyces coelicolor* A3(2) depends on MreB for proper localization and contributes to spore wall synthesis. J Bacteriol 193:1533–1542

Henry JT, Crosson S (2013) Chromosome replication and segregation govern the biogenesis and inheritance of inorganic polyphosphate granules. Mol Biol Cell 24:3177–3186

Hill NS, Buske PJ, Shi Y, Levin PA (2013) A moonlighting enzyme links *Escherichia coli* cell size with central metabolism. PLoS Genet 9:e1003663

Hill NS, Kadoya R, Chattoraj DK, Levin PA (2012) Cell size and the initiation of DNA replication in bacteria. PLoS Genet 8:e1002549

Hiraoka S, Matsuzaki H, Shibuya I (1993) Active increase in cardiolipin synthesis in the stationary growth phase and its physiological significance in *Escherichia coli*. FEBS Lett 336:221–224

Hottes AK, Shapiro L, McAdams HH (2005) DnaA coordinates replication initiation and cell cycle transcription in Caulobacter crescentus. Mol Microbiol 58:1340–1353

Hou Z, Bernstein DA, Fox CA, Keck JL (2005) Structural basis of the Sir1-origin recognition complex interaction in transcriptional silencing. Proc Natl Acad Sci USA 102:8489–8494

Hsu HC, Stillman B, Xu RM (2005) Structural basis for origin recognition complex 1 protein-silence information regulator 1 protein interaction in epigenetic silencing. Proc Natl Acad Sci U S A 102:8519–8524

Huijbregts RP, Svitin A, Stinnett MW, Renfrow MB, Chesnokov I (2009) Drosophila Orc6 facilitates GTPase activity and filament formation of the septin complex. Mol Biol Cell 20:270–281

Ingerson-Mahar M, Gitai Z (2012) A growing family: the expanding universe of the bacterial cytoskeleton. FEMS Microbiol Rev 36:256–266

Ishida T, Akimitsu N, Kashioka T, Hatano M, Kubota T, Ogata Y, Sekimizu K, Katayama T (2004) DiaA, a novel DnaA-binding protein, ensures the timely initiation of *Escherichia coli* chromosome replication. J Biol Chem 279:45546–45555

Ishikawa S, Ogura Y, Yoshimura M, Okumura H, Cho E, Kawai Y, Kurokawa K, Oshima T, Ogasawara N (2007) Distribution of stable DnaA-binding sites on the Bacillus subtilis genome detected using a modified ChIP-chip method. DNA Res 14:155–168

James R (1975) Identification of an outermembrane protein of *Escherichia coli*, with a role in the coordination of deoxyribonucleic acid replication and cell elongation. J Bacteriol 124:918–929

Jannière L, Canceill D, Suski C, Kanga S, Dalmais B, Lestini R, Monnier AF, Chapuis J, Bolotin A, Titok M, Le Chatelier E, Ehrlich SD (2007) Genetic evidence for a link between glycolysis and DNA replication. PLoS One 2:e447

Jin DJ, Cagliero C, Zhou YN (2013) Role of RNA Polymerase and transcription in the organization of the bacterial nucleoid. Chem Rev 113:8662–8682

Kaguni JM (2011) Replication initiation at the *Escherichia coli* chromosomal origin. Curr Opin Chem Biol 15:606–613

Kang S, Han JS, Park JH, Skarstad K, Hwang DS (2003) SeqA protein stimulates the relaxing and decatenating activities of topoisomerase IV. J Biol Chem 278:48779–48785

Kasho K, Katayama T (2013) DnaA binding locus datA promotes DnaA-ATP hydrolysis to enable cell cycle-coordinated replication initiation. Proc Natl Acad Sci U S A 110:936–941

Katayama T, Ozaki S, Keyamura K, Fujimitsu K (2010) Regulation of the replication cycle: conserved and diverse regulatory systems for DnaA and oriC. Nat Rev Microbiol 8:163–170

Kawai Y, Daniel RA, Errington J (2009) Regulation of cell wall morphogenesis in Bacillus subtilis by recruitment of PBP1 to the MreB helix. Mol Microbiol 71:1131–1144

Kawakami H, Ozaki S, Suzuki S, Nakamura K, Senriuchi T, Su'etsugu M, Fujimitsu K, Katayama T (2006) The exceptionally tight affinity of DnaA for ATP/ADP requires a unique aspartic acid residue in the AAA+ sensor 1 motif. Mol Microbiol 62:1310–1324

Kawakami H, Keyamura K, Katayama T (2005) Formation of an ATP-DnaA-specific initiation complex requires DnaA Arginine 285, a conserved motif in the AAA+ protein family. J Biol Chem 280:27420–27430

Kawakami H, Iwura T, Takata M, Sekimizu K, Hiraga S, Katayama T (2001) Arrest of cell division and nucleoid partition by genetic alterations in the sliding clamp of the replicase and in DnaA. Mol Genet Genomics 266:167–179

Keener J, Nomura M (1996) Regulation of ribosome synthesis. In: Neidhardt FC (ed) *Escherichia coli* and Salmonella,1. ASM Press, Washington, pp 1417–1431

Kelly AJ, Sackett MJ, Din N, Quardokus E, Brun YV (1998) Cell cycle-dependent transcriptional and proteolytic regulation of FtsZ in Caulobacter. Genes Dev 12:880–893

Képès F, Jester BC, Lepage T, Rafiei N, Rosu B, Junier I (2012) The layout of a bacterial genome. FEBS Lett 586:2043–2048

Keyamura K, Fujikawa N, Ishida T, Ozaki S, Su'etsugu M, Fujimitsu K, Kagawa W, Yokoyama S, Kurumizaka H, Katayama T (2007) The interaction of DiaA and DnaA regulates the replication cycle in E. coli by directly promoting ATP DnaA-specific initiation complexes. Genes Dev 21:2083–2099

Kim JW, Dang CV (2005) Multifaceted roles of glycolytic enzymes. Trends Biochem Sci 30:142–150

Kitchen JL, Li Z, Crooke E (1999) Electrostatic interactions during acidic phospholipid reactivation of DnaA protein, the *Escherichia coli* initiator of chromosomal replication. Biochemistry 38:6213–6221

Klungsøyr HK, Skarstad K (2004) Positive supercoiling is generated in the presence of *Escherichia coli* SeqA protein. Mol Microbiol 54:123–131

Krasny L, Gourse RL (2004) An alternative strategy for bacterial ribosome synthesis: *Bacillus subtilis* rRNA transcription regulation. EMBO J 23:4473–4483

Kuhlman TE, Cox EC (2012) Gene location and DNA density determine transcription factor distributions in *Escherichia coli*. Mol Syst Biol 8:610

Kühner S, van Noort V, Betts MJ, Leo-Macias A, Batisse C, Rode M, Yamada T, Maier T, Bader S, Beltran-Alvarez P, Castaño-Diez D, Chen WH, Devos D, Güell M, Kurokawa K, Nishida S, Emoto A, Sekimizu K, Katayama T (1999) Replication cycle-coordinated change of the adenine nucleotide-bound forms of DnaA protein in *Escherichia coli*. EMBO J 18:6642–6652

Kurth I, Georgescu RE, O'Donnell ME (2013) A solution to release twisted DNA during chromosome replication by coupled DNA polymerases. Nature 496:119–122

Lavelle C (2014) Pack, unpack, bend, twist, pull, push: the physical side of gene expression. Curr Opin Genet Dev 25C:74–84

Le TB, Imakaev MV, Mirny LA, Laub MT (2013) High-resolution mapping of the spatial organization of a bacterial chromosome. Science 342:731–734

Lempiäinen H, Shore D (2009) Growth control and ribosome biogenesis. Curr Opin Cell Biol 21:855–863

Leng F, Chen B, Dunlap DD (2011) Dividing a supercoiled DNA molecule into two independent topological domains. Proc Natl Acad Sci USA 108:19973–19978

Leng F, McMacken R (2002) Potent stimulation of transcription-coupled DNA supercoiling by sequence-specific DNA-binding proteins. Proc Natl Acad Sci U S A 99:9139–9144

Leonard AC, Grimwade JE (2011) Regulation of DnaA assembly and activity: taking directions from the genome. Annu Rev Microbiol 65:19–35

Leonard AC, Grimwade JE (2010) Regulating DnaA complex assembly: it is time to fill the gaps. Curr Opin Microbiol 13:766–772

Levine A, Vannier F, Dehbi M, Henckes G, Seror SJ (1991) The stringent response blocks DNA replication outside the *ori* region in *Bacillus subtilis* and at the origin in *Escherichia coli*. J Mol Biol 219:605–613

Li Z, Kitchen JL, Boeneman K, Anand P, Crooke E (2005) Restoration of growth to acidic phospholipid-deficient cells by DnaA(L366K) is independent of its capacity for nucleotide binding and exchange and requires DnaA. J Biol Chem 280:9796–9801

Lin YL, Pasero P (2012) Interference between DNA replication and transcription as a cause of genomic instability. Curr Genomics 13:65–73

Liu LF, Wang JC (1987) Supercoiling of the DNA template during transcription. Proc Natl Acad Sci U S A 84:7024–7027

Liu F (2014) Qimuge, Hao J, Yan H, Bach T, Fan L, Morigen. AspC-Mediated Aspartate Metabolism Coordinates the *Escherichia coli* Cell Cycle. PLoS One 9:e92229

Løbner-Olesen A, Slominska-Wojewodzka M, Hansen FG, Marinus MG (2008) DnaC inactivation in *Escherichia coli* K-12 induces the SOS response and expression of nucleotide biosynthesis genes. PLoS One 3:e2984

Løbner-Olesen A, Marinus MG, Hansen FG (2003) Role of SeqA and Dam in *Escherichia coli* gene expression: a global/microarray analysis. Proc Natl Acad Sci U S A 100:4672–4677

Løbner-Olesen A, Boye E (1992) Different effects of mioC transcription on initiation of chromosomal and minichromosomal replication in *Escherichia coli*. Nucleic Acids Res 20:3029–3036

Loeb LA, Springgate CF, Battula N (1974) Errors in DNA replication as a basis of malignant changes. Cancer Res 34:2311–2321

Loeb LA, Loeb KR, Anderson JP (2003) Multiple mutations and cancer. Proc Natl Acad Sci USA 100:776–781

MacAlpine HK, Gordân R, Powell SK, Hartemink AJ, MacAlpine DM (2010) Drosophila ORC localizes to open chromatin and marks sites of cohesin complex loading. Genome Res 20:201–211

Machné R, Murray DB (2012) The yin and yang of yeast transcription: elements of a global feedback system between metabolism and chromatin. PLoS One 7:e37906

Maciąg M, Kochanowska M, Łyżeń R, Węgrzyn G, Szalewska-Pałasz A (2010) ppGpp inhibits the activity of *Escherichia coli* DnaG primase. Plasmid 63:61–67

Maciąg M, Nowicki D, Janniere L, Szalewska-Pałasz A, Węgrzyn G (2011) Genetic response to metabolic fluctuations: correlation between central carbon metabolism and DNA replication in *Escherichia coli*. Microb Cell Fact 10:19

Maciąg M, Nowicki D, Szalewska-Pałasz A, Węgrzyn G (2012) Central carbon metabolism influences fidelity of DNA replication in *Escherichia coli*. Mutat Res 731:99–106

Maciąg-Dorszyńska M, Ignatowska M, Jannière L, Węgrzyn G, Szalewska-Pałasz A (2012) Mutations in central carbon metabolism genes suppress defects in nucleoid position and cell division of replication mutants in *Escherichia coli*. Gene 503:31–35

Maciąg-Dorszyńska M, Szalewska-Pałasz A, Węgrzyn G (2013) Different effects of ppGpp on *Escherichia coli* DNA replication in vivo and in vitro. FEBS Open Bio 3:161–164

Makise M, Mima S, Tsuchiya T, Mizushima T (2000) Identification of amino acids involved in the functional interaction between DnaA protein and acidic phospholipids. J Biol Chem 275:4513–4518

Margulies C, Kaguni JM (1996) Ordered and sequential binding of DnaA protein to oriC, the chromosomal origin of *Escherichia coli*. J Biol Chem 271:17035–17040

Marr AG, Ingraham JL (1962) Effect of temperature on the composition of fatty acids in *Escherichia coli*. J Bacteriol 84:1260–1267

Marsh D (2008) Protein modulation of lipids, and vice-versa, in membranes. Biochim Biophys Acta 1778:1545–1575

Mathews CK (2006) DNA precursor metabolism and genomic stability. FASEB J 20:1300–1314

McGarry KC, Ryan VT, Grimwade JE, Leonard AC (2004) Two discriminatory binding sites in the *Escherichia coli* replication origin are required for DNA strand opening by initiator DnaA-ATP. Proc Natl Acad Sci U S A 101:2811

Messer W (2002) The bacterial replication initiator DnaA. DnaA and oriC, the bacterial mode to initiate DNA replication. FEMS Microbiol Rev 26:355–374

Messer W, Blaesing F, Jakimowicz D, Krause M, Majka J, Nardmann J, Schaper S, Seitz H, Speck C, Weigel C, Wegrzyn G, Welzeck M, Zakrzewska-Czerwinska J (2001) Bacterial replication initiator DnaA. Rules for DnaA binding and roles of DnaA in origin unwinding and helicase loading. Biochimie 83:5–12

Messer W, Weigel C (2003) DnaA as a transcription regulator. Methods Enzymol 370:338–349

Michelsen O (2003) Teixeira de Mattos MJ, Jensen PR, Hansen FG. Precise determinations of C and D periods by flow cytometry in *Escherichia coli* K-12 and B/r. Microbiology 149:1001–1010

Mijakovic I, Petranovic D, Macek B, Cepo T, Mann M, Davies J, Jensen PR, Vujaklija D (2006) Bacterial single-stranded DNA-binding proteins are phosphorylated on tyrosine. Nucleic Acids Res 34:1588–1596

Mileykovskaya E, Dowhan W (2009) Cardiolipin membrane domains in prokaryotes and eukaryotes. Biochim Biophys Acta 1788:2084–2091

Miller DT, Grimwade JE, Betteridge T, Rozgaja T, Torgue JJ, Leonard AC (2009) Bacterial origin recognition complexes direct assembly of higher-order DnaA oligomeric structures. Proc Natl Acad Sci U S A 106:18479–18484

Montero Llopis P, Jackson AF, Sliusarenko O, Surovtsev I, Heinritz J, Emonet T, et al. (2010) Spatial organization of the flow of genetic information in bacteria. Nature. 466: 77–81

Molina F, Skarstad K (2004) Replication fork and SeqA focus distributions in *Escherichia coli* suggest a replication hyperstructure dependent on nucleotide metabolism. Mol Microbiol 52:1597–1612

Morigen, Odsbu I, Skarstad K (2009) Growth rate dependent numbers of SeqA structures organize the multiple replication forks in rapidly growing *Escherichia coli*. Genes Cells 14: 643–57

Mozharov AD, Shchipakin VN, Fishov IL (1985) Evtodienko YuV. Changes in the composition of membrane phospholipids during the cell cycle of *Escherichia coli*. FEBS Lett 186:103–106

Muskhelishvili G, Travers A (2013) Integration of syntactic and semantic properties of the DNA code reveals chromosomes as thermodynamic machines converting energy into information. Cell Mol Life Sci 70:4555–4567

Nasmyth K, Haering CH (2009) Cohesin: its roles and mechanisms. Annu Rev Genet 43:525–558

Newman G, Crooke E (2000) DnaA, the initiator of *Escherichia coli* chromosomal replication, is located at the cell membrane. J Bacteriol 182:2604–2610

Nishida S, Fujimitsu K, Sekimizu K, Ohmura T, Ueda T, Katayama T (2002) A nucleotide switch in the *Escherichia coli* DnaA protein initiates chromosomal replication: evidnece from a mutant DnaA protein defective in regulatory ATP hydrolysis in vitro and in vivo. J Biol Chem 277:14986–14995

Noirot-Gros MF, Dervyn E, Wu LJ, Mervelet P, Errington J, Ehrlich SD, Noirot P (2002) An expanded view of bacterial DNA replication. Proc Natl Acad Sci USA 99:8342–8347

Norambuena T, Racke I, Rybin V, Schmidt A, Yus E, Aebersold R, Herrmann R, Böttcher B, Frangakis AS, Russell RB, Serrano L, Bork P, Gavin AC (2009) Proteome organization in a genome-reduced bacterium. Science 326:1235–1240

Norregaard K, Andersson M, Sneppen K, Nielsen PE, Brown S, Oddershede LB (2013) DNA supercoiling enhances cooperativity and efficiency of an epigenetic switch. Proc Natl Acad Sci U S A 110:17386–17391

Norris V, Amar P (2012) Life on the scales. Life 2:286–312

Norris V, Nana GG, Audinot JN (2013) New approaches to the problem of generating coherent, reproducible phenotypes. Theory Biosci 133:47–61

Nowosielska A, Nieminuszczy J, Grzesiuk E (2004) Spontaneous mutagenesis in exponentially growing and stationary-phase, *umuDC*-proficient and -deficient, *Escherichia coli dnaQ49*. Acta Biochim Pol 54:683–692

Nozaki S, Ogawa T (2008) Determination of the minimum domain II size of *Escherichia coli* DnaA protein essential for cell viability. Microbiology 154:3379–3384

Nozaki S, Yamada Y, Ogawa T (2009) Initiator titration complex formed at datA with the aid of IHF regulates replication timing in *Escherichia coli*. Genes Cells 14:329–341

Olliver A, Saggioro C, Herrick J, Sclavi B (2010) DnaA-ATP acts as a molecular switch to control levels of ribonucleotide reductase expression in *Escherichia coli*. Mol Microbiol 76:1555–1571

Omont N, Képès F (2004) Transcription/replication collisions cause bacterial transcription units to be longer on the leading strand of replication. Bioinformatics 20:2719–2725

Ote T, Hashimoto M, Ikeuchi Y, Su'etsugu M, Suzuki T, Katayama T, Kato J (2006) Involvement of the *Escherichia coli* folate-binding protein YgfZ in RNA modification and regulation of chromosomal replication initiation. Mol Microbiol 59:265–275

Ozaki S, Katayama T (2012) Highly organized DnaA-oriC complexes recruit the single-stranded DNA for replication initiation. Nucleic Acids Res 40:1648–1665

Ozaki S, Katayama T (2009) DnaA structure, function, and dynamics in the initiation at the chromosomal origin. Plasmid 62:71–82

Ozaki S, Noguchi Y, Hayashi Y, Miyazaki E, Katayama T (2012) Differentiation of the DnaA-oriC subcomplex for DNA unwinding in a replication initiation complex. J Biol Chem 287:37458–37471

Ozaki T, Kumaki Y, Kitagawa R, Ogawa T (2001) Anomalous DnaA protein binding to the regulatory region of the *Escherichia coli* aldA gene. Microbiology 147:153–159

Ozaydin B, Rine J (2010) Expanded roles of the origin recognition complex in the architecture and function of silenced chromatin in Saccharomyces cerevisiae. Mol Cell Biol 30:626–639

Pak DT, Pflumm M, Chesnokov I, Huang DW, Kellum R, Marr J, Romanowski P, Botchan MR (1997) Association of the origin recognition complex with heterochromatin and HP1 in higher eukaryotes. Cell 91:311–323

Peter BJ, Arsuaga J, Breier AM, Khodursky AB, Brown PO, Cozzarelli NR (2004) Genomic transcriptional response to loss of chromosomal supercoiling in *Escherichia coli*. Genome Biol 5:R87

Pflumm MF, Botchan MR (2001) Orc mutants arrest in metaphase with abnormally condensed chromosomes. Development 128:1697–1707

Phillips R, Ursell T, Wiggins P, Sens P (2009) Emerging roles for lipids in shaping membrane-protein function. Nature 459:379–385

Polaczek P, Wright A (1990) Regulation of expression of the dnaA gene in *Escherichia coli*: role of the two promoters and the DnaA box. New Biol 2:574–582

Pomerantz RT, O'Donnell M (2010) What happens when replication and transcription complexes collide? Cell Cycle 9:2537–2543

Potrykus K, Murphy H, Philippe N, Cashel M (2011) ppGpp is the major source of growth rate control in E. coli. Environ Microbiol 13: 563–575

Potrykus K, Cashel M (2008) (p)ppGpp: still magical? Annu Rev Microbiol 62:35–51

Prasanth SG, Shen Z, Prasanth KV, Stillman B (2010) Human origin recognition complex is essential for HP1 binding to chromatin and heterochromatin organization. Proc Natl Acad Sci U S A 107:15093–15098

Prasanth SG, Prasanth KV, Stillman B (2002) Orc6 involved in DNA replication, chromosome segregation, and cytokinesis. Science 297:1026–1031

Preston BD, Albertson TM, Herr AJ (2010) DNA replication fidelity and cancer. Semin Cancer Biol 20:281–293

Quardokus E, Din N, Brun YV (1996) Cell cycle regulation and cell type-specific localization of the FtsZ division initiation protein in Caulobacter. Proc Natl Acad Sci U S A 93:6314–6319

Radhakrishnan SK, Pritchard S, Viollier PH (2010) Coupling prokaryotic cell fate and division control with a bifunctional and oscillating oxidoreductase homolog. Dev Cell 18:90–101

Ramachandran L, Burhans DT, Laun P, Wang J, Liang P, Weinberger M, Wissing S, Jarolim S, Suter B, Madeo F, Breitenbach M, Burhans WC (2006) Evidence for ORC-dependent repression of budding yeast genes induced by starvation and other stresses. FEMS Yeast Res 6:763–776

Ramamurthi KS, Lecuyer S, Stone HA, Losick R (2009) Geometric cue for protein localization in a bacterium. Science 323:1354–1357

Ramamurthi KS, Losick R (2009) Negative membrane curvature as a cue for subcellular localization of a bacterial protein. Proc Natl Acad Sci USA 106:13541–13545

Ray PS, Arif A, Fox PL (2007) Macromolecular complexes as depots for releasable regulatory proteins. Trends Biochem Sci 32:158–164

Rocha EP (2004) The replication-related organization of bacterial genomes. Microbiology 150:1609–1627

Roth A, Messer W (1995) The DNA binding domain of the initiator protein DnaA. EMBO J 14:2106–2111

Ryan VT, Grimwade JE, Camara JE, Crooke E, Leonard AC (2004) *Escherichia coli* prereplication complex assembly is regulated by dynamic interplay among Fis. IHF and DnaA Mol Microbiol 51:1347–1359

Scholefield G, Murray H (2013) YabA and DnaD inhibit helix assembly of the DNA replication initiation protein DnaA. Mol Microbiol 90:147–159

Regev T, Myers N, Zarivach R, Fishov I (2012) Association of the chromosome replication initiator DnaA with the *Escherichia coli* inner membrane in vivo: quantity and mode of binding. PLoS One 7:e36441

Rocha EP (2008) Evolutionary patterns in prokaryotic genomes. Curr Opin Microbiol 11:454–460

Rocha EP, Danchin A (2003) Gene essentiality determines chromosome organisation in bacteria. Nucleic Acids Res 31:6570–6577

Ross W, Vrentas CE, Sanchez-Vazquez P, Gaal T, Gourse RL (2013) The magic spot: a ppGpp binding site on E. coli RNA polymerase responsible for regulation of transcription initiation. Mol Cell 50:420–429

Roth A, Urmoneit B, Messer W (1994) Functions of histone-like proteins in the initiation of DNA replication at oriC of *Escherichia coli*. Biochimie 76:917–923

Rozgaja TA, Grimwade JE, Iqbal M, Czerwonka C, Vora M, Leonard AC (2011) Two oppositely oriented arrays of low-affinity recognition sites in oriC guide progressive binding of DnaA during *Escherichia coli* pre-RC assembly. Mol Microbiol 82:475–488

Rudner DZ, Losick R (2010) Protein subcellular localization in bacteria. Cold Spring Harb Perspect Biol 2:a000307

Rymer RU, Solorio FA, Tehranchi AK, Chu C, Corn JE, Keck JL et al (2012) Binding mechanism of metal NTP substrates and stringent response alarmones to bacterial DnaG-type primases. Structure 20:1478–1489

Saberi S, Emberly E (2010) Chromosome driven spatial patterning of proteins in bacteria. PLoS Comput Biol 6:e1000986

Saggioro C, Olliver A, Sclavi B (2013) Temperature-dependence of the DnaA-DNA interaction and its effect on the autoregulation of dnaA expression. Biochem J 449:333–341

Sánchez-Romero MA, Molina F, Jiménez-Sánchez A (2010) Correlation between ribonucleoside-diphosphate reductase and three replication proteins in *Escherichia coli*. BMC Mol Biol 11:11

Sasaki T, Gilbert DM (2007) The many faces of the origin recognition complex. Curr Opin Cell Biol 19:337–343

Saxena R, Fingland N, Patil D, Sharma AK, Crooke E (2013) Crosstalk between DnaA Protein, the Initiator of *Escherichia coli* Chromosomal Replication, and Acidic Phospholipids Present in Bacterial Membranes. Int J Mol Sci 14:8517–8537

Saxena R, Rozgaja T, Grimwade J, Crooke E (2011) Remodeling of nucleoprotein complexes is independent of the nucleotide state of a mutant AAA+ protein. J Biol Chem 286:33770–33777

Schaper S, Messer W (1995) Interaction of the initiator protein DnaA of *Escherichia coli* with its DNA target. J Biol Chem 270(29):17622

Scholefield G, Veening JW, Murray H (2011) DnaA and ORC: more than DNA replication initiators. Trends Cell Biol 21:188–194

Seitz H, Weigel C, Messer W (2000) The interaction domains of the DnaA and DnaB replication proteins of *Escherichia coli*. Mol Microbiol 37:1270–1279

Shapiro L, McAdams HH, Losick R (2009) Why and how bacteria localize proteins. Science 326:1225–1228

Sherwood R, Takahashi TS, Jallepalli PV (2010) Sister acts: coordinating DNA replication and cohesion establishment. Genes Dev 24:2723–2731

Shi IY, Stansbury J, Kuzminov A (2005) A defect in the acetyl coenzyme A<–> acetate pathway poisons recombinational repair-deficient mutants of *Escherichia coli*. J Bacteriol 187:1266–1275

Shimada K, Gasser SM (2007) The origin recognition complex functions in sister-chromatid cohesion in Saccharomyces cerevisiae. Cell 128:85–99

Shor E, Warren CL, Tietjen J, Hou Z, Müller U, Alborelli I, Gohard FH, Yemm AI, Borisov L, Broach JR, Weinreich M, Nieduszynski CA, Ansari AZ, Fox CA (2009) The origin recognition complex interacts with a subset of metabolic genes tightly linked to origins of replication. PLoS Genet 5:e1000755

Simmons LA, Felczak M, Kaguni JM (2003) DnaA Protein of *Escherichia coli*: oligomerization at the E. coli chromosomal origin is required for initiation and involves specific N-terminal amino acids. Mol Microbiol 49:849–858

Singh D, Chang SJ, Lin PH, Averina OV, Kaberdin VR, Lin-Chao S (2009) Regulation of ribonuclease E activity by the L4 ribosomal protein of *Escherichia coli*. Proc Natl Acad Sci U S A 106:864–869

Sirover MA (2011) On the functional diversity of glyceraldehyde-3-phosphate dehydrogenase: biochemical mechanisms and regulatory control. Biochim Biophys Acta 1810:741–751

Skarstad K, Torheim N, Wold S, Lurz R, Messer W, Fossum S, Bach T (2001) The *Escherichia coli* SeqA protein binds specifically to two sites in fully and hemimethylated oriC and has the capacity to inhibit DNA replication and affect chromosome topology. Biochimie 83:49–51

Skarstad K, Katayama T (2013) Regulating DNA replication in bacteria. Cold Spring Harb Perspect Biol 5:a012922

Slominińska M, Wegrzyn A, Konopa G, Skarstad K, Wegrzyn G (2001) SeqA, the *Escherichia coli* origin sequestration protein, is also a specific transcription factor. Mol Microbiol 40:1371–1379

Słominińska M, Konopa G, Ostrowska J, Kedzierska B, Wegrzyn G, Wegrzyn A (2003) SeqA-mediated stimulation of a promoter activity by facilitating functions of a transcription activator. Mol Microbiol 47:1669–1679

Smelkova N, Marians KJ (2001) Timely release of both replication forks from oriC requires modulation of origin topology. J Biol Chem 276:39186–39191

Sobetzko P, Glinkowska M, Travers A, Muskhelishvili G (2013) DNA thermodynamic stability and supercoil dynamics determine the gene expression program during the bacterial growth cycle. Mol Biosyst 9:1643–1651

Sobetzko P, Travers A, Muskhelishvili G (2012) Gene order and chromosome dynamics coordinate spatiotemporal gene expression during the bacterial growth cycle. Proc Natl Acad Sci USA, 109: E42–E50

Speck C, Messer W (2001) Mechanism of origin unwinding: sequential binding of DnaA to double- and single-stranded DNA. EMBO J 20:1469–1476

Speck C, Weigel C, Messer W (1999) ATP- and ADP-dnaA protein, a molecular switch in gene regulation. EMBO J 18:6169–6176

Spitzer J (2011) From water and ions to crowded biomacromolecules: in vivo structuring of a prokaryotic cell. Microbiol Mol Biol Rev 75:491–506

Stein A, Firshein W (2000) Probable identification of a membrane-associated repressor of Bacillus subtilis DNA replication as the E2 subunit of the pyruvate dehydrogenase complex. J Bacteriol 182:2119–2124

Stepankiw N, Kaidow A, Boye E, Bates D (2009) The right half of the *Escherichia coli* replication origin is not essential for viability, but facilitates multi-forked replication. Mol Microbiol 74:467–479

Strauss B, Kelly K, Dincman T, Ekiert D, Biesieda T, Somg R (2004) Cell death in *Escherichia coli dnaE*(Ts) mutants incubated at a nonpermissive temperature is prevented by mutation in the *cydA* gene. J Bacteriol 186:2147–2155

Su'etsugu M, Emoto A, Fujimitsu K, Keyamura K, Katayama T (2003) Transcriptional control for initiation of chromosomal replication in *Escherichia coli*: fluctuation of the level of origin transcription ensures timely initiation. Genes Cells 8:731–745

Suter B, Tong A, Chang M, Yu L, Brown GW, Boone C, Rine J (2004) The origin recognition complex links replication, sister chromatid cohesion and transcriptional silencing in Saccharomyces cerevisiae. Genetics 167:579–591

Sutton MD, Carr KM, Vicente M, Kaguni JM (1998) *Escherichia coli* DnaA protein. The N-terminal domain and loading of DnaB helicase at the E. coli chromosomal origin. J Biol Chem 273:34255–34262

Szalewska-Pałasz A, Wegrzyn A, Herman A, Wegrzyn G (1994) The mechanism of the stringent control of lambda plasmid DNA replication. EMBO J 13:5779–5785

Szambowska A, Pierechod M, Wegrzyn G, Glinkowska M (2011) Coupling of transcription and replication machineries in λ DNA replication initiation: evidence for direct interaction of *Escherichia coli* RNA polymerase and the λO protein. Nucleic Acids Res 39:168–177

Tabata S, Oka A, Sugimoto K, Takanami M, Yasuda S, Hirota Y (1983) The 245 base-pair oriC sequence of the E. coli chromosome directs bidirectional replication at an adjacent region. Nucleic Acids Res 11:2617–2626

Takahashi TS, Yiu P, Chou MF, Gygi S, Walter JC (2004) Recruitment of Xenopus Scc2 and cohesin to chromatin requires the pre-replication complex. Nat Cell Biol 6:991–996

Teplyakov A, Obmolova G, Sarikaya E, Pullalarevu S, Krajewski W, Galkin A, Howard AJ, Herzberg O, Gilliland GL (2004) Crystal structure of the YgfZ protein from *Escherichia coli* suggests a folate-dependent regulatory role in one-carbon metabolism. J Bacteriol 186:7134–7140

Theisen PW, Grimwade JE, Leonard AC, Bogan JA, Helmstetter CE (1993) Correlation of gene transcription with the time of initiation of chromosome replication in *Escherichia coli*. Mol Microbiol 10:575–584

Tojo S, Satomura T, Kumamoto K, Hirooka K, Fujita Y (2008) Molecular mechanism underlying the positive stringent response of the *Bacillus subtilis ilv-leu* operon, involved in the biosynthesis of branched-chain amino acids. J Bacteriol 190:6143–6147

Travers A, Muskhelishvili G (2005) DNA supercoiling—a global transcriptional regulator for enterobacterial growth? Nat Rev Microbiol 3:157–169

Ulbrich B, Nierhaus KH (1975) Pools of ribosomal proteins in *Escherichia coli*. Studies on the exchange of proteins between pools and ribosomes. Eur J Biochem 57:49–54

van Workum M, van Dooren SJ, Oldenburg N, Molenaar D, Jensen PR, Snoep JL, Westerhoff HV (1996) DNA supercoiling depends on the phosphorylation potential in *Escherichia coli*. Mol Microbiol 20:351–360

Vanounou S, Parola AH, Fishov I (2003) Phosphatidylethanolamine and phosphatidylglycerol are segregated into different domains in bacterial membrane. A study with pyrene-labelled phospholipids. Mol Microbiol 49:1067–1079

Vanounou S, Pines D, Pines E, Parola AH, Fishov I (2002) Coexistence of domains with distinct order and polarity in fluid bacterial membranes. Photochem Photobiol 76:1–11

Veening JW, Murray H, Errington J (2009) A mechanism for cell cycle regulation of sporulation initiation in Bacillus subtilis. Genes Dev 23:1959–1970

Vendeville A, Larivière D, Fourmentin E (2011) An inventory of the bacterial macromolecular components and their spatial organization. FEMS Microbiol Rev 35:395–414

Versalovic J, Lupski JR (1997) Missense mutations in the 3′ end of the Escherichia coli dnaG gene do not abolish primase activity but do confer the chromosomesegregation-defective (par) phenotype. Microbiology 143:585–594

von Meyenburg K, Boye E, Skarstad K, Koppes L, Kogoma T (1987) Mode of initiation of constitutive stable DNA replication in RNase H-defective mutants of Escherichia coli K-12. J Bacteriol 169:2650–2658

Waldminghaus T, Weigel C, Skarstad K (2012) Replication fork movement and methylation govern SeqA binding to the Escherichia coli chromosome. Nucleic Acids Res 40:5465–5476

Waldminghaus T, Skarstad K (2009) The Escherichia coli SeqA protein. Plasmid 61:141–150

Wang JD, Levin PA (2009) Metabolism, cell growth and the bacterial cell cycle. Nat Rev Microbiol 7:822–827

Wang JD, Sanders GM, Grossman AD (2007) Nutritional control of elongation of DNA replication by (p)ppGpp. Cell 128:865–875

Wang X (2013) Montero Llopis P, Rudner DZ. Organization and segregation of bacterial chromosomes. Nat Rev Genet 14:191–203

Warner JR, McIntosh KB (2009) How common are extraribosomal functions of ribosomal proteins? Mol Cell 34:3–11

Weart RB, Lee AH, Chien AC, Haeusser DP, Hill NS, Levin PA (2007) A metabolic sensor governing cell size in bacteria. Cell 130:335–347

Węgrzyn G, Licznerska K, Węgrzyn A (2012) Phage λ—new insights into regulatory circuits. Adv Virus Res 82:155–178

Wegrzyn G, Wegrzyn A (2005) Genetic switches during bacteriophage lambda development. Prog Nucleic Acid Res Mol Biol 79:1–48

Weigel C, Schmidt A, Seitz H, Tüngler D, Welzeck M, Messer W (1999) The N-terminus promotes oligomerization of the Escherichia coli initiator protein DnaA. Mol Microbiol 34:53–66

Werner JN, Chen EY, Guberman JM, Zippilli AR, Irgon JJ, Gitai Z (2009) Quantitative genome-scale analysis of protein localization in an asymmetric bacterium. Proc Natl Acad Sci U S A 106:7858–7863

Wheeler LJ, Rajagopal I, Mathews CK (2005) Stimulation of mutagenesis by proportional deoxyribonucleoside triphosphate accumulation in Escherichia coli. DNA Repair 4:1450–1456

White CL, Kitich A, Gober JW (2010) Positioning cell wall synthetic complexes by the bacterial morphogenetic proteins MreB and MreD. Mol Microbiol 76:616–633

Williams JS, Smith DJ, Marjavaara L, Lujan SA, Chabes A, Kunkel TA (2013) Topoisomerase 1-mediated removal of ribonucleotides from nascent leading-strand DNA. Mol Cell 49:1010–1015

Wold S, Skarstad K, Steen HB, Stokke T, Boye E (1994) The initiation mass for DNA replication in Escherichia coli K-12 is dependent on growth rate. EMBO J 13:2097–2102

Wolfe AJ (2010) Physiologically relevant small phosphodonors link metabolism to signal transduction. Curr Opin Microbiol 13:204–209

Wróbel B, Murphy H, Cashel M, Wegrzyn G (1998) Guanosine tetraphosphate (ppGpp)-mediated inhibition of the activity of the bacteriophage lambda pR promoter in Escherichia coli. Mol Gen Genet 257:490–495

Xia W, Dowhan W (1995) In vivo evidence for the involvement of anionic phospholipids in initiation of DNA replication in Escherichia coli. Proc Natl Acad Sci U S A 92:783–787

Yamaguchi Y, Hase M, Makise M, Mima S, Yoshimi T, Ishikawa Y, Tsuchiya T, Mizushima T (1999) Involvement of Arg-328, Arg-334 and Arg-342 of DnaA protein in the functional interaction with acidic phospholipids. Biochem J 340:433–438

Yu H, Dröge P (2014). Replication-induced supercoiling: a neglected DNA transaction regulator? Trends Biochem Sci (in press), doi:10.1016/j.tibs.2014.02.009

Yu FX, Dai RP, Goh SR, Zheng L, Luo Y (2009) Logic of a mammalian metabolic cycle: an oscillated NAD+/NADH redox signaling regulates coordinated histone expression and S-phase progression. Cell Cycle 8:773–779

Yung BY, Kornberg A (1988) Membrane attachment activates dnaA protein, the initiation protein of chromosome replication in *Escherichia coli*. Proc Natl Acad Sci U S A 85:7202–7205

Zhang Y, Yu Z, Fu X, Liang C (2002) Noc3p, a bHLH protein, plays an integral role in the initiation of DNA replication in budding yeast. Cell 109:849–860

Zheng L, Roeder RG, Luo Y (2003) S phase activation of the histone H2B promoter by OCA-S, a coactivator complex that contains GAPDH as a key component. Cell 114:255–266

Zheng W, Li Z, Skarstad K, Crooke E (2001) Mutations in DnaA protein suppress the growth arrest of acidic phospholipid-deficient *Escherichia coli* cells. EMBO J 20:1164–1172

Zundel MA, Basturea GN, Deutscher MP (2009) Initiation of ribosome degradation during starvation in *Escherichia coli*. RNA 15:977–983

Zuo Y, Wang Y, Steitz TA (2013) The mechanism of *E. coli* RNA polymerase regulation by ppGpp is suggested by the structure of their complex. Mol Cell 50:430–436

Zyskind JW, Cleary JM, Brusilow WS, Harding NE, Smith DW (1983) Chromosomal replication origin from the marine bacterium Vibrio harveyi functions in *Escherichia coli*: oriC consensus sequence. Proc Natl Acad Sci U S A 80:1164–1168